FUELS FROM WASTE

ENERGY SCIENCE AND ENGINEERING:
RESOURCES, TECHNOLOGY, MANAGEMENT
An International Series
EDITOR
JESSE DENTON
Belton, Texas

LARRY L. ANDERSON and DAVID A. TILLMAN (eds.), Fuels from Waste, 1977

In preparation

A. J. ELLIS and W. A. J. MAHON, Chemistry and Geothermal Systems
FRANCIS G. SHINSKEY, Energy Conservation through Control

FUELS FROM WASTE

Edited by

LARRY L. ANDERSON

University of Utah
Salt Lake City, Utah

DAVID A. TILLMAN

Materials Associates, Inc.
Washington, D.C.

ACADEMIC PRESS   New York   San Francisco   London   1977
A Subsidiary of Harcourt Brace Jovanovich, Publishers

ACADEMIC PRESS, INC.
111 Fifth Avenue, New York, New York 10003

United Kingdom Edition published by
ACADEMIC PRESS, INC. (LONDON) LTD.
24/28 Oval Road, London NW1

Library of Congress Cataloging in Publication Data

Main entry under title:

Fuels from waste.

    Includes bibliographical references.
    1.   Waste products as fuel.   I.   Anderson, Larry
LaVon.   II.   Tillman, David A.
TP360.F83   662'.6   77-74023
ISBN 0−12−056450−5

For Sharon and Millie

CONTENTS

Chapter IV   **Pipeline Gas from Solid Wastes by the Syngas Recycling Process**

Herman F. Feldmann, G. W. Felton, H. Nack, and J. Adlerstein

Chapter V   **The Nature of Pyrolytic Oil from Municipal Solid Waste**

Kenneth W. Pober and H. Fred Bauer

Chapter VI   **The Conversion of Feedlot Wastes into Pipeline Gas**

Frederick T. Varani and John J. Burford, Jr.

Chapter VII   **Fuels and Chemicals from Crops**

Henry R. Bungay and Roscoe F. Ward

LIST OF CONTRIBUTORS

Numbers in parentheses indicate the pages on which the authors' contributions begin.

J. ADLERSTEIN (57), Syngas Recycling Corporation, West Toronto, Ontario, Canada

LARRY L. ANDERSON (1), Department of Fuels Engineering, University of Utah, Salt Lake City, Utah

HERBERT R. APPELL (121), Pittsburgh Energy Research Center, Energy Research and Development Administration, Pittsburgh, Pennsylvania

K. M. BARCLAY (41), Atomics International Division, Rockwell International Corporation, Canoga Park, California

H. FRED BAUER (73), Occidental Research Corporation, La Verne, California

HENRY R. BUNGAY* (105), Division of Solar Energy, Energy Research and Development Administration, Washington, D.C.

JOHN J. BURFORD, JR.† (87), Research Division, Bio-Gas of Colorado, Inc., Arvada, Colorado

HERMAN F. FELDMANN (57), Battelle-Columbus Laboratories, Columbus, Ohio

G. W. FELTON (57), Battelle Columbus Laboratories, Columbus, Ohio

G. HAIDER (171), Department of Fuels Engineering, University of Utah, Salt Lake City, Utah

K. T. LIU (161), Battelle-Columbus Laboratories, Columbus, Ohio

H. NACK (57, 161), Battelle-Columbus Laboratories, Columbus, Ohio

* Present address: Department of Chemical and Environmental Engineering, Rensselaer Polytechnic Institute, Troy, New York.
† Present address: Bio-Gas of Colorado, Inc., Loveland, Colorado.

J. H. OXLEY (161), Battelle-Columbus Laboratories, Columbus, Ohio

KENNETH W. POBER (73), Occidental Research Corporation, La Verne, California

FRED SHAFIZADEH (141), Department of Chemistry, University of Montana, Missoula, Montana

G. RAY SMITHSON, JR. (195), Environmental Technology Program Office, Battelle-Columbus Laboratories, Columbus, Ohio

E. P. STAMBAUGH (161), Battelle-Columbus Laboratories, Columbus, Ohio

DAVID A. TILLMAN (17, 211), Materials Associates, Inc., Washington, D.C.

FREDERICK T. VARANI (87), Research Division, Bio-Gas of Colorado, Inc., Arvada, Colorado

ROSCOE F. WARD (105), Division of Solar Energy, Energy Research and Development Administration, Washington, D.C.

S. J. YOSIM (41), Atomics International Division, Rockwell International Corporation, Canago Park, California

PREFACE

The $25+ billion bill which this country pays annually for imported oil provides one measure of the problem. The $6+ billion we pay for municipal waste disposal provides another measure of the problem. This is a society which is short on energy and long on rubbish, garbage, woodwaste, manure, and other residues of the production and consumption processes.

The synergistic proposition of creating gold from garbage, or fuels from waste in this case, is well known. The processes and products associated with accomplishing the task have been defined less precisely. The concept is sound. The specifics merit close consideration.

This book addresses the following questions: How much waste is being generated? Which systems are presently available to produce fuels from waste? Which technologies are being developed for converting municipal, agricultural, silvicultural, and industrial solid wastes into fuels? Which nontechnical problems must be solved if residues are to become a more significant energy source?

The volume focuses on the production of fuels rather than energy. That distinction is important. The fuel product desired determines the nature of the waste utilization technology. The character of any specific fuel determines its potential market and its market potential. The chemical composition of any fuel determines its synergy with existing fossil fuels— hence its degree of user acceptance. While the problems of market potential and user acceptance appear to be nontechnical in orientation, they can be solved only by technology.

After the volume and composition of organic wastes generated and their relationship to energy availability are defined, the book focuses on those technologies resulting in useful fuels. While a distinction is made between presently available and future technologies, the current state of the art for research systems is not shunned. Good processes now exist. Too often our optimism about any given process is directly proportional to the distance (in time) of that technology from commercial application. Existing processes, such as fluidized-bed combustion, merit the same consideration as future processes, such as wood waste liquefaction. Both present and future processes merit careful technical consideration.

Following the consideration of the technical issues, the nontechnical problems of product marketing and system financing are surveyed. These are economic aspects of the problem which engineers and scientists must consider in the design of new technologies for the conversion of wastes into useful fuels. All projects, when successful, marry technology and economics. The solutions of problems result from such interweaving of disciplines.

To achieve this focused examination of fuels from waste, the editors relied on and obtained excellent cooperation from the many contributors. They bring to the book not only their specific written contributions but also many excellent concepts concerning the direction which this assessment should take. The editors also gained valuable assistance from Dr. Bernard Blaustein of the Pittsburgh Energy Research Center (ERDA), Dr. Wendell Wiser of the University of Utah, Dr. Joseph H. Oxley of Battelle Memorial Laboratories, and Dr. James Boyd and Dr. Earl Hayes of Materials Associates, Inc.

What began as a concentrated symposium at the April 1976 American Chemical Society Meetings is now a far broader and more complete examination for your consideration.

Chapter I

A WEALTH OF WASTE; A SHORTAGE OF ENERGY

Larry L. Anderson
DEPARTMENT OF FUELS ENGINEERING
UNIVERSITY OF UTAH
SALT LAKE CITY, UTAH

I. INTRODUCTION

A. Energy Sources and Waste

The basic energy sources are those available to the earth's surface environment from basically different sources. These consist of the following four categories:

(1) gravitational (from the earth–moon–sun system),
(2) nuclear reactions (natural or artificial fission or fusion),
(3) geothermal (from the earth's interior), and
(4) solar radiation.

All of these energy sources are being considered for more extensive development and use by mankind. Some power plants are being operated which convert the kinetic energy of the tides into electricity [1]. Nuclear fission is also being used in many parts of the world to generate electric power. The power plants that utilize fission reactions will probably supply a growing percentage of the total electric power in the coming few decades. The most serious delays currently affecting the growth of the nuclear power industry are associated with the environmental effects of the by-products of fission and with thermal addition to the environment.

Nuclear fusion may provide a long-term answer to the energy resource problems but although some 50 fusion experiments have been conducted none has yielded more energy than it has consumed. There are still several serious engineering problems to be solved before atomic fusion can become a viable energy source, if indeed it can be. Geothermal energy is being developed and utilized as particular conditions permit. In some limited areas geothermal heat and/or steam or water can contribute significant quantities of energy (the Geysers in northern California; the Salton Sea in southern California). In terms of the total energy requirements for the United States or other large countries gravitational and geothermal energy sources are not expected to constitute major portions.

Solar energy which reaches the earth's atmosphere is largely reflected back into space as short or long wave radiation. The major part of that which is not reflected is used to heat the atmosphere and the earth and to drive the hydrologic cycle (evaporation, precipitation, and runoff of surface water). A very small fraction of the incident solar radiation is utilized by living organisms primarily to produce plant tissue by photosynthesis. The plant material produced by photosynthesis is mostly returned to the earth and the atmosphere by decay and oxidation. The fossil fuels represent only a very small amount of stored plant material which was retained and concentrated by special conditions of entrapment in the earth's crust.

Because these materials were submerged out of the oxidizing environment of the atmosphere their rate of decay was greatly reduced. Pressure, temperature, and time have resulted in the accumulation of these materials in differing forms such as coal, petroleum, oil shale, tar sands, natural gas, and other related materials. The storage process which has resulted in the accumulation of the fossil fuels could be much more efficient. For example, it has been estimated that the total organic carbon production of the earth by photosynthesis is about 150×10^9 tons/year [2]. The oxidation of only this carbon would yield about 4200×10^{15} BTU or over 100 times the world's estimated energy requirements in the year 2000. It should be emphasized that this estimate is for new carbon which would be in addition to existing plants not grown new each year. This of course ignores the hydrogen and other elements in the plant material which could also yield energy. Actually this oxidation is taking place all the time but the heat released is low-temperature heat and not too useful for many of the needs of man in an industrialized society.

Table I summarizes the energy flow into the earth's surface environment [3].

B. Plant Matter as Waste

As stated previously the amount of energy utilized in the production of plant material represents only a small fraction of the solar radiation reaching the earth. The fraction of the plant matter actually utilized by men and animals in our industrialized society is even smaller. However, the quantity

TABLE I

Flow of Energy into the Environment

Inputs	Forms	Magnitude (10^{12} W)	Approximate percentage of the total
Gravitational	Tidal currents	3.0	0.002
Nuclear reactions	Heat	0.05	
Fossil fuels	Heat, chemicals	5.8	0.005
Geothermal	Lava, hot springs steam, hot rock	35.0	0.020
Solar radiation	Heat	81,000	66.692
	Winds, currents, waves, etc.	370	0.305
	Hydrologic cycle	40,000	32.934
	Photosynthesis	40	0.033
			99.991

of this matter is not insignificant as evidenced by the environmental problems created by the disposal of only the waste from the use of such plant
(and animal) material. The variety of organic matter that is termed waste is
long and varied and ranges from waste that is of little consequence to the
environment to that which not only pollutes but may spread disease and
death.

The solid waste problems of the United States have received much attention in recent years and an attempt will not be made here to discuss these in
detail. However, the types and quantities of these wastes which are organic
(or contain organic matter) will be explored along with their potential as
energy sources.

C. Generation of Organic Waste

Every segment of any society produces solid wastes containing organic
matter. In sparsely populated regions this may not represent an environmental problem. However, in an industrialized society with large cities,
industrial centers, and agricultural production areas accumulation can
become overwhelming. The following sections review the current and
projected estimates of the most important waste materials.

II. QUANTITIES AND SOURCES OF ORGANIC WASTES

The most abundant waste materials containing organic solids are manure,
urban refuse, and agricultural wastes. The total of these various organic
wastes generated yearly in the United States is well over 2 billion tons. The
largest portion of this waste is manure and other agricultural wastes. Urban
refuse (including domestic, municipal, and commercial), sewage solids, and
industrial wastes also constitute a significant portion. Industrial wastes are
defined here as those resulting from an industrial operation or establishment, not collected as part of the urban refuse. Logging and wood wastes
form the final category for consideration.

The quantities of organic waste generated in the United States have been
given by several authors during the past several years. The basis of calculating these quantities has not always been uniform. In the following sections
the quantities calculated are "dry organic solids." In many publications by
authors in agriculture the term "volatile solids" is used. This is defined as
the solid material volatilized in a furnace at 650°C [4]. This term has been
assumed to by synonymous with "dry organic solids," that is, moisture- and
ash-free organic material.

A. Manure

The total quantity of animal wastes has been given in several ways. Manure or animal waste production has been described to be "in excess of" 1.7 billion tons [5], approximately 1.7 billion tons [6], 2 billion tons [7-9], or 1,562,721,000 tons [10].

These quantities describing the amount of "manure" produced by various farm animals may represent different things. Loehr has reviewed this problem in detail as follows [8]:

> A review of a large number of publications reveals that the term "manure" may mean any one of a number of things: (1) fresh excrement including both the solid and liquid portions, (2) total excrement but with enough bedding added to absorb the liquid portion, (3) the remaining part of the total excrement after most of the liquid has drained away, (4) the remaining material after liquid drainage, evaporation of water, and leaching of soluble nutrients, or (5) only the liquid which has been allowed to drain from the total excrement. The water content of each of the preceding materials is highly variable.

The important point is that most of the values given for "manure" or "animal wastes" include water ranging from small percentages to 87%. When evaluating these materials as energy sources, the water must be excluded. The organic solid materials which can be converted to oil or gas for fuel turn out to be much smaller quantities than those previously described.

Minimum quantities of animal wastes can be calculated from data available on the physical and chemical characteristics of animal wastes, the quantities produced by each animal species, and animal population statistics. Loehr [8], Miner [5], and Taiganides and Hazen [11] have all tabulated data on fecal waste production by various kinds of livestock. Table II summarizes the waste generation by major farm animals. Estimating how much waste animals generate is complicated by the fact that waste production may vary significantly depending on the conditions of confinement, type of feed, and other factors. The data in the table given by Loehr take this into account, and this consideration results in a range of values for dry "volatile solids" [8]. A reasonable current minimum value of the total organic solids produced from major farm animals as shown in Table II is 210 million tons/year. A maximum value for organic wastes from this source, based on characteristics of animal manures and total quantities generated, is approximately 250 million tons. As a waste disposal problem, however, the amount of wet manure produced and therefore in need of disposal or processing is approximately 1.7 billion tons, as previously stated.

TABLE II

Wastes Generated by Major Farm Animals, 1976

Animals	Number of animals in the United States 1976[a] (thousands)	Daily wastes (lb/animal)		Organic solids/year (moisture- and ash-free)	
		Solids	Organic solids	Tons/animal	For all animals (million tons)[b]
Cattle	127,976	10.25 (9.5–11.4)[c]	8.20 (7.6–9.1)	1.50 (1.39–1.66)	192.0 (178–214)
Hogs	49,602	1.2 (0.8–1.6)	0.91 (0.68–1.36)	0.166 (0.124–0.248)	8.2 (6.2–12.3)
Sheep	13,346	0.56	0.46	0.084	1.1
Poultry	1,036,832[d]	0.06 (0.05–0.10)	0.047 (0.04–0.077)	0.0086 (0.0073–0.0141)	8.9 (7.57–14.6)
Total					210.0 (192–242)

[a] USDA [12].

[b] Figures in parentheses are ranges which are values given by Loehr [8] and account for the fact that the amount of manure produced by animals is dependent on such factors as confinement conditions, type of feed, and size of animal.

[c] Assuming average live weight of cattle, 1000 lb; hogs, 167 lb; sheep, 67 lb; and chickens, 3.44 lb [15].

[d] Includes all hens, pullets, breeder hens, turkeys, and total number of broilers, divided by 4.5, which corrects for the life span of only a few months for broilers.

B. Agricultural Crop Wastes

The generation of field wastes by major agricultural crops is high. The solid wastes generated by major agricultural crops in 1966 was estimated to be 550 million tons. The trend has been, in recent years, that each year more food and agricultural crops are grown on a smaller land area. The wastes have been increasing, but at a very low rate. Cornstalks, pea vines, sugarcane stalks, leaves, stubble, prunings, and similar wastes from other plants constitute most of this waste. Much of this material is now burned to prevent the spread of plant disease or is left on farmland to prevent wind and water erosion. Since these wastes are, to a large extent, not available at centralized locations, they are doubtful sources for conversion to other energy forms. There are some exceptions, however, such as bagasse from sugar mills, stalks and cobs of corn, and some milling wastes from wheat, rice, and other grains. Because of increased mechanization and constantly improving methods of harvesting crops, less waste is being taken from the fields to central processing plants. This trend will continue as further improvements are made. It is estimated that 70% of the solid crop wastes are organic solids and that this amounts to 400 million tons of organic solids a year.

C. Urban Refuse

Urban refuse is being generated constantly. The domestic, municipal, and commercial components of this waste amount to 3.6, 1.2, and 2.3 lb per capita per day, respectively. The total amount of this waste in 1971 was estimated at 260 million tons and in 1976 at 330 million tons. At least half of this is dry organic material, and some analyses claim that many as 80% of municipal wastes are carbonaceous [13–15]. At the present time approximately 90% of this refuse is disposed of in landfills, while most of the remainder is burned in municipal incinerators. The quantity of refuse per capita is expected to increase in the future. Landfilling and incineration are now being attacked as methods of disposal since pollution of either groundwater or air results while available space for landfills is running out and costs are increasing. Conversion of urban refuse to oil or gas for fuel and generation of electric power by incineration would appear to be a more suitable method of disposal. The energy source represented by this waste is constantly renewable. The quantity of organic solids currently generated from urban refuse amounts to 165 million tons/year, based on analyses presented in the references previously cited. Other reports have estimated that municipal solid waste will increase by up to 15% on a per capita basis and by 25% of the 1980 total municipal waste [16].

D. Sewage Solids

Municipal sewage contains organic solids which must be filtered, treated, or removed if the water is to be recycled. Increasing pressures to bring this about are becoming more evident. Sewage, however, is principally water, about 99.8% [17], with the remaining materials in suspension or solution. Organic materials in sewage include human wastes, paper, and food scraps. Gates has further classified the organic materials in domestic sewage (exclusive of ground garbage) as 50% carbohydrates, 40% nitrogenous matter, and 10% fats [18]. In 1976 approximately 20 million tons of organic solids were generated as suspended and dissolved solids (sewage sludge) based on 0.5 lb of organic solids per capita per day.

E. Industrial Wastes

Industrial wastes are any materials discarded from industrial operations. These include processing, packaging, shipping, office, and other wastes. Since industries vary so much in the type and quantity of materials handled and processed, the composition of the wastes from one industry may be entirely different from that of another. Sludges, waste plastic, rags, paper and cardboard, scrap metals, slag, rubber, ceramics, etc., may be in the waste from a single industrial operation. The waste generated by industry has been estimated to be 150 million tons annually. Since the quantity and composition of this waste vary from one operation to another, the contained organic material will be available only at particular locations. There is also some difficulty in calculating a reasonable value for the quantity of organic materials present in industrial wastes. However, from data accumulated from several sources, 40% seems a reasonable minimum [14]. On this basis some 60 million tons of organic waste solids are generated by industry annually. It should be emphasized that these wastes do not include mineral wastes, which are estimated to amount to 1.7 billion tons annually and are expected to be well over 2 billion tons by 1980 [19,10].

F. Logging and Wood Manufacturing Residues

Approximately one-third of the volume of wood harvested in the United States is unused. This is a significant industrial or quasi-industrial waste which, because of volume, is its own category. This amounted to 4.4 billion ft^3 of wood in 1968, of which 2 billion ft^3 were logging residues. The amount of wood wastes will probably remain constant or decrease in the next few decades even though the demand for both softwood and hardwood products will continue to increase. This is primarily due to the diligence with which the timber industry utilizes its waste materials. The most serious problem is bark which comprises about one-fourth of the primary manufacturing wastes (those resulting from processing at sawmills, veneer mills, and pulp mills) [20]. Some of these wastes could be a source of raw material for conversion to electric power or to oil or gas for fuel. Much of the logging waste left in forests is not concentrated, but, due to its potential fire hazard and its potential for spreading tree diseases, it may be collected and become available. At present the logging and wood processing residues amount to 80 million tons/year. This, of course, does not include the wood fiber-based portion of household wastes (45–50% of household wastes is wood or wood products).

G. Miscellaneous Organic Wastes

In addition to wastes shown in previous categories, there are many others which, when added together, constitute a significant quantity. These include antibiotic fermentation residues, wastes and carcasses from cats, dogs, horses, and marine animals, organic wastes from federal installations, and many other wastes not previously accounted for. A conservative estimate for the quantity of this materials is 80 million tons/year.

III. ENERGY POTENTIAL FROM ORGANIC WASTES, 1976 AND 1980

The total quantity of organic wastes generated in the United States in 1976 is summarized in Table III. Estimates of the quantities that will be

TABLE III

Estimates of Organic Wastes Generated, 1976 and 1980[a]

Source	1976	1980
Manure	210	260
Agricultural crops and food wastes[b]	400	400
Urban refuse	165	220
Municipal sewage solids	20	25
Industrial wastes[c]	60	76
Logging and wood manufacturing residues	80	80
Miscellaneous organic wastes	80	90
Total	1015	1151

    [a] 10^6 tons/year.
    [b] Assuming 70% dry organic solids in major agricultural crop waste solids.
    [c] Based on 150 million tons of industrial wastes per year in 1976.

generated in 1980 are also given. The values are somewhat conservative since some specific sources of organic wastes are omitted because data were not available. The quantities given for 1980 are based on a population of 226 million. Some increases, which are not caused by population increases, are expected. These include an annual increase in the per capita consumption of beef from 112 lb in 1970 to 130 lb in 1980. The amount of urban refuse per capita is also expected to increase to at least 8 lb/day, a figure which represents only collected wastes. Although larger quantities of agricultural crops will be produced in 1980, agricultural crop waste figures for that year were not increased because there is no conclusive proof that crop waste quantity is directly related to crop yield. The data given in Table III update estimates made in the past for 1971 [21] and 1974 [22].

IV. AVAILABILITY OF ORGANIC WASTES AS ENERGY SOURCES

The quantities of organic wastes given in Table III are potential but include much material that is not collected and is not likely to be collected in the foreseeable future. If a commercial process for the conversion of these organic wastes to usable energy is developed, the feed materials for the processes discussed in other chapters of this book will then be quantities of waste that constitute a fuel or energy source. Successful processing of wastes would stimulate the collection of other wastes and increase the amount of these materials available for processing.

An estimate of the quantity of organic waste which has accumulated and which would be available as an energy source is described in the sections that follow. This estimate is a minimum value and includes only quantities that are now concentrated by municipal collection, feedlot operation, and the like.

A. Manure from Cattle, Chickens, and Hogs

Of the 210 million tons of dry organic solids from farm animals generated annually, estimates indicate that 50 to 80% comes from animals raised in confined conditions, that is, feedlots, dairies, egg laying, turkey farms, or broiler operations. The trend in the United States is to fewer but larger operations of this type. The waste accumulation from beef cattle makes up most of that generated by farm animals in confinement. The data on cattle feedlots show that not only are more animals being confined to feedlots with 1000 or more head per lot but that the fastest growing feedlots are the largest ones (those with 32,000 or more head per lot). Similar trends are evident on poultry farms where 100,000 to 1,000,000 birds per operation are increasing, on dairy farms where several hundred to more than 1000 cows per farm are common, and in hog-feeding operations where several thousand hogs may be confined at a single location.

In some areas such as Texas, Kansas, Nebraska, and Iowa, several large feedlots may be located close enough for accumulation of manure in a single location for processing. The distance the manure can reasonably be transported is not great; however, removal and disposal of most manure from feedlots involve some expense. Some manure may be processed for refeeding or for crop fertilizer, but this does not appear to be an adequate solution.

Assuming that only animal wastes from feedlots having 1000 or more cattle and those from the largest poultry and hog operations can be considered as energy sources, it is estimated that approximately 30 million tons of dry organic solids would be available. If successful processing were established, the size of feedlots and their proximity to one another would be affected, which could make more manure available for processing.

B. Agricultural Crop Wastes

Although the tonnage of crop waste and crop spoilage is significant much of this material is widely dispersed and is usually disposed of on the farm. The practice of burning harvest residues may be limited in the future, transforming more of this material into a waste disposal problem. At the present time the organic material available from this source for energy is limited but significant in size at specific processing plants such as canneries, mills, slaughterhouses, and dairies. Assuming that only corn milling and canning, slaughtering of animals, sugar refining, and dairying have quantities of waste available for use as energy sources, the amount of this source is estimated at 116 million tons. Allowing for nearly 80% water, the organic waste still amounts to 25 million tons.

C. Urban Refuse

Landfilling is the most predominant method for disposal of urban refuse. In the United States it is estimated that 90% of the urban refuse that is collected is disposed of in this manner at some 12,000 individual sites. Incineration accounts for most of the remaining solid waste disposal (about 8% of the total). However, many problems exist in incineration technology and there is some question whether conventional incineration plants can meet the air quality standards now being imposed in many cities. The amount of refuse generated in the 100 largest population centers in the United States in 1971 was about 160 million tons, which included 70 million tons of dry organic solids. This estimate is based on the population of urban areas and assumes the same composition of urban refuse as previously cited. The organic material from these most populous areas can be easily collected and could constitute a significant energy source. Other data indicate that the situation for collection may improve with time. For example, the annual recovery of paper and combustible matter has been estimated at 77 million tons (with 57.7% recovery) by 1985 [16].

D. Sewage Solids

Because of increased concern of the public over the pollution of rivers, streams, lakes, and even the oceans, the usual methods for disposing of municipal sewage have become obsolete, and in some cases illegal. The principal pollutant in this sewage is organic matter. The availability of this organic matter for conversion to oil will depend somewhat on the legal requirements imposed in municipalities to treat sewage water before disposal or recycle. A conservative estimate for 1976 is that 2.0 million tons of organic solids are available from this source. This number could easily become much greater since sewage systems are already essentially "processing" systems which could be tapped for the usable materials without significant alteration.

E. Industrial Wastes

The amount of industrial wastes collected has been estimated at 23% of the total amount generated. Considering the amount of organic material in industrial wastes, 7.0 million tons of organic solids are collected and disposed of from industry and can be considered as available.

F. Logging and Wood Manufacturing Wastes

Logging and wood manufacturing residues from timber products in the United States were estimated at 4.4 billion ft^3 (55×10^6 tons) in 1968. Forty-five percent of this material (2.0 billion ft^3) was left in the forests as logging debris. Although this material is a reservoir for plant diseases and a serious fire hazard, it will probably not be collected or concentrated unless a profit can be realized by its transportation to a processing location. Because of the projected demands for increased timber products, most of the wood wastes should, in fact, decrease because of improved forest management practices and more efficient utilization of the wood actually processed at mills. Undoubtedly, some bark, sawdust, and other materials will continue to accumulate at specific locations and will require disposal. The amount of this material now available is estimated at 5.0 million tons, most of which is now disposed of by burning.

G. Available Organic Wastes

Table IV summarizes estimates of organic wastes currently available and which present a disposal problem since they are concentrated at locations where disposal requires some special expense, processing step, or transportation. These estimates are somewhat conservative since they neglect to include many specific organic wastes such as manure from dairies, several major crop wastes, food and other scraps disposed of in sewers, paunch manure, dead animals and their parts not collected at major slaughtering plants, fermentation wastes, and many other sources of organic materials. Miscellaneous organic wastes are assumed at 10% of the total amount generated.

Large quantities of these available wastes are now concentrated at specific locations. For example, a single cattle feedlot with 100,000 head would produce about 410 tons of dry organic solids a day or 150,000 tons/ year. A city and its nearby suburbs with a population of 1 million would generate 1750 tons of dry organic solids daily or 640,000 tons/year. Naturally at a feedlot or city of these sizes, the amount of waste material to be handled would be considerably more than the amounts given since the organic material is only part of the waste. There are several areas where a large city or several cities could collect their wastes in a common location. There are also locations in the high plains of West Texas and in California where several large feedlots or dairies are in close proximity.

In actual practice, a city and its suburbs could combine its urban refuse, sewage sludge, industrial wastes, manure, and agricultural wastes. Alternatively, these wastes could be added to conventional fuels used for generating

TABLE IV

Estimates of Available Organic Wastes, 1976[a]

Source	Total organic wastes generated	Organic solids available
Manure	210	30
Urban refuse	165	70
Logging and wood manufacturing residues	80	5
Agricultural crops and food wastes	400	25
Industrial wastes	60	7
Municipal sewage solids	20	2
Miscellaneous organic wastes	80	8
Total	1015	147

[a] Millions of tons per year.

TABLE V

Typical Analyses for Fuel Materials

	Peat	Wood (dried)	Wood bark (dried Douglas fir)	Dry bagasse	Municipal refuse (Altoona, Pennsylvania)	Paper mill sludge (dried)	Bituminous coal West (Utah)	Bituminous coal East (Pittsburgh seam)	Heat oil #2
Proximate Analysis (%)									
Moisture	91.0			35.0	43.3	23.2	5.0	3.0	0.1
Volatile matter	5.4	81.5	73.0		43.0		47.6	33.9	
Fixed carbon	3.0	17.5	26.0		6.7		48.3	55.8	
Ash	1.6	1.0	1.3		7.0	10.2	4.1	10.3	
Ultimate analysis (%)									
Hydrogen	5.7	6.3	5.9	6.1	8.2	7.2	6.0	5.0	14.2
Carbon	58.0	52.0	56.2	47.3	27.2	30.9	77.9	75.5	85.0
Nitrogen	1.2	0.1			0.7	0.5	1.5	1.2	
Oxygen	35.0	40.5	36.7	35.3	56.8	51.2	9.9	4.9	
Sulfur	0.11		trace		0.1	0.2	0.6	3.1	1.0
Ash	1.6	1.0		11.3	7.0	10.2	4.1	10.3	
Heating value (BTU/lb of refuse)									
Moist	1,200		3,000	4,870	4,830	5,350	13,470	13,250	20,400
Dried		9,000	9,500	9,140			14,170	13,650	20,400

electric power. Their effect would be to lower the sulfur content of the conventional fuel. Several of the following chapters discuss the advantages of particular methods of using these wastes to make available more useful forms of energy.

V. CONCLUSION

Waste materials as energy sources are much like conventional fuels. They vary in composition, density, heating value, and other properties. The value of a particular fuel or waste as an energy resource will depend on several factors, including the availability of large quantities near potential markets. The composition of the waste or fuel is certainly one of the most important factors related to its value. Table V gives some typical analyses and heating values for wastes and fuels. In almost every case other values could have been selected since each of these materials is only representative.

The information given in Table V shows that wastes contain significant organic material which can be burned or processed in a manner similar to conventional fuels. The composition of wastes is similar in many ways to such fuels as coal, except that wastes generally have higher oxygen and moisture content. On the other hand, wastes generally have low sulfur content and many have lower ash content than typical coals.

The quantity of waste materials generated annually in the United States is enormous and on the increase. Many of these materials contain organic or combustible solids and liquids. The collection and disposal of these wastes are becoming more expensive and difficult. Already several processes are being used either to burn the organic material in wastes or to convert it to more concentrated forms. This is being done in process development units, pilot plants, and, in a few cases, in full-scale plants. It seems only logical that organic wastes will become more valuable as energy sources in the future. Population centers will find utilization of these materials more attractive than the present "reverse mining" process of landfilling at an ever increasing expense. Industrial, commercial, and agricultural sectors will also continue to realize the value of wastes. If this takes place, a thriving and permanent industry will develop involving the combustion and/or conversion of wastes to useful energy. The following chapters describe some of the ways this will be achieved as well as some of the problems, both technical and institutional, which will be encountered.

REFERENCES

1. T. J. Healy, "Energy, Electric Power and Man," Chap. 8, pp. 177–192. Boyd & Fraser, Publ. Co., San Francisco, California, 1974.

2. G. A. Riley, The carbon metabolism and photo-synthetic efficiency of the earth as a whole, *Am. Sci.* **32**(2), 129–134 (1944).

3. M. K. Hubbert, The energy resources of the earth, *Sci. Am.* **224**(3), 60–70, (1971).

4. S. A. Hart, Fowl fecal facts, *World's Poult. Sci. J.* **19**(4), 263–272 (1963).

5. J. R. Miner, Farm animal-waste management. Iowa State Univ. of Science and Technology, Agric. and Home Ec. Expt. Station, Ames, Iowa, Special Rep. 67, May 1971.

6. Control of agricultural related pollution. Rep. Submitted by Sec. of Agriculture and Dir. of the Office of Science and Techonology, U.S. Dept. of Agriculture, January 1969.

7. C. E. Knapp, Agriculture poses waste problems, *Environ. Sci. Technol.* **4**(12), 1098–1100, (1970).

8. R. C. Loehr, Pollution implications of animal wastes—a forward oriented review. Federal Water Pollution Control Administration, Robert S. Kerr Water Research Center, Ada, Oklahoma, July 1968.

9. C. H. Wadleigh, Wastes in relation to agriculture and forestry. U.S. Department of Agriculture, Misc. Pub. 1065, March 1968.

10. Solid waste management, a comprehensive assessment of solid waste problems, practices, and needs. Executive Office of the President, Ad Hoc Group for Office of Science and Technology, May 1969.

11. E. P. Taiganides and T. E. Hazen, Properties of farm animal excreta, *Trans. Am. Soc. Agric. Eng.* **9**(3), 374–376, (1966).

12. Statistical reporting service crop reporting board 1976 livestock and poultry inventory. U.S. Dept. of Agriculture, January 1976.

13. B. H. Rosen, R. G. Evans, P. Carabelli, and R. B. Zaborowski, Economic evaluation of a commercial size refuse pyrolysis plant. Cities Service Oil Co., Rep. 1, March 1970.

14. K. C. Dean, C. J. Chindgren, and L. Peterson, Preliminary separation of metals and nonmetals from urban refuse. U.S. Bur. Mines, Tech. Prog. Rep. 34, June 1971.

15. W. S. Sanner, C. Ortuglio, J. G. Walters, and D. E. Wolfson, Conversion of municipal and industrial refuse into useful materials by pyrolysis. U.S. Bur. Mines, Rep. of Investigations 7428, 1970.

16. W. C. Franklin, Potential for resource recovery in the United States. Rep. Prepared for the Aluminum Co. of America, March 1975.

17. Development of a coal-based sewage treatment process. U.S. Office of Coal Research, Final Rep., Res. and Development Rep. 55, p. 13, 1971.

18. C. D. Gates, The disposal of domestic wastes in rural areas, *in* Agriculture and the Quality of Our Environment, pp. 367–384. Am. Assoc. for the Advancement of Science, Pub. 85, 1967.

19. Material needs and the environment today and tomorrow. Final Rep. of the National Commission on Materials Policy, 4E-3, 4E-5, June 1973.

20. E. P. Cliff, Timber: the renewable material, pp. 3–6, Table 3.2. Prepared for the National Commission on Materials Policy, August 1973.

21. L. L. Anderson, Energy potential from organic wastes, a review of the quantities and sources. U.S. Bur. Mines, Information Circular 8549, 1972.

22. I. Wender, F. W. Steffgen, and P. M. Yavorsky, Clean liquid and gaseous fuels from organic solid wastes, *in* "Recycling and Disposal of Solid Wastes (T. F. Yen, ed.), Chap. 2, pp. 43–99. Ann Arbor Science Publishers, Inc., Ann Arbor, Michigan, 1974.

Chapter II

ENERGY FROM WASTES: AN OVERVIEW OF PRESENT TECHNOLOGIES AND PROGRAMS

David A. Tillman
MATERIALS ASSOCIATES, INC.
WASHINGTON, D.C.

I. INTRODUCTION

Producing fuel from wastes accomplishes two essential tasks: it reduces the existence of environmentally hazardous situations and it increases the supply of energy produced from indigenous resources. Both problems must be addressed by society. Thus, the National Commission on Materials Policy recommended in its letter of transmittal to the President and the Congress: "Conserve our natural resources and environment by treating waste materials as resources and returning them either to use or, in a harmless condition, to the ecosystems" [1].

As was noted in Chapter I, the United States now generates 165 million tons of municipal solid waste (MSW) annually. That waste, when interred into the ground, creates numerous hazards. "Reverse mines" or landfills generate potentially explosive methane and produce toxic and polluting leachates which contaminate water systems. On the other side of the coin, the United States continues to import growing amounts of energy from unstable sources.

Both waste disposal and energy availability are increasingly costly problems. The annual national expenditures for collecting and landfilling 165 million tons of MSW have surpassed $6 million [2]. Additional economic losses result from the misplacement of these wastes. Virgin sources of energy—petroleum, natural gas, and coal—must be extracted or imported while waste is discarded. Conservatively, this increases the annual balance of payments outflow to meet the need for additional energy by $3 billion [3]. That loss of dollars could be staunched by converting organic wastes into useful energy.

The environmental and economic costs associated with the energy from waste issue necessitate a review of present technology for converting waste to energy. Such a review must include a redefinition of solid waste plus a discussion of the techniques available for using MSW, a uniform comparison of those technologies, and a consideration of the rate at which they are being adopted by communities and regions.

II. WASTE AS AN ENERGY ORE

Municipal solid waste can be viewed not only as worthless trash but also as "urban ore." In the latter context, it is both a low grade and an excellent ore. It is a low grade ore because the high value constituents appear in low concentrations. Solid waste contains less than 10% ferrous metals, whereas taconites contain more than 30% iron. Urban ore contains less than 1% aluminum, whereas bauxite contains over 50% alumina—over 25% aluminum. Solid waste contains from 5000 to 6000 BTU/lb whereas bituminous coal possesses 11,000 BTU/lb. Solid waste, however, remains an excellent ore. It appears much like native ore because contained values such as iron are discrete elements. It has a low density and a high surface area-to-mass ratio; thus, valuable fractions can be liberated easily because they enjoy a high level of exposure within the mass of ore. Finally, its value stems from the fact that, although heterogeneous, over 80% of this ore is virtually always useful. Many elements of the ore which are impurities from an energy product viewpoint, ferrous and nonferrous metals for example, may be removed from the energy fraction and sold as products in their own right. Table I presents the composition of "urban ore."

The cursory description of urban ore composition in Table I provides an initial examination of the heterogeneity of this matter while demonstrating that organic rather than metallic materials must dominate resource recovery programs. Two examples from the organic waste stream demonstrate that the ore is even more complex than the initial view suggests. Paper, for example, contains large and increasing volumes of printing inks, adhesives, plastics, resins, varnishes, clays, latex asphalt, waxes, and gums [4]. Plastics in waste include low density polyethylene, high density polyethylene, polystyrene, polyvinyl chloride (PVC), polypropylene, phenolics, polyester, alkyds, epoxies, melamine, and urethane foam [5]. Such complexities virtually preclude complete segregation and recovery of these organics as materials and mandate their use as energy, except in special situations. For this reason, the concept of "urban ore" can be refined further; solid waste can be considered as an extremely young coal.

Mine run coal appears to be a carbon-based mineral conglomerate containing varying amounts of carbon, hydrogen, sulfur, alumina, iron, and a host of other substances. Similarly, urban waste contains varying amounts of mineral constituents. Unlike coal, however, the organic components of MSW contain varying amounts of energy, as Table II demonstrates.

The comparisons between coal and waste extend beyond the analysis of

TABLE I

Estimate of the Composition of Municipal Solid Waste[a]

Waste component	Percentage of the total
Organic materials	73.6
Paper products	42.0
Food wastes	12.0
Yard wastes	15.0
Plastics, rubber	1.6
Textiles	0.6
Wood	2.4
Inorganic materials	26.4
Ferrous and nonferrous metals	8.0
Glass	6.0
Rocks and dirt	11.0
Other	1.4

[a] Source: National Solid Waste Management Association.

TABLE II

Energy Values in Municipal Waste[a]

Component	Average heat value (as-received) (BTU/lb)
Paper and cardboard	8000
Food waste	2500
Yard waste	3000
Wood	7000
Textiles	7000
Leather and rubber	7000
Plastics	15,000
Average	6000

[a] Source: Pittsburgh Energy Research Center.

constituents. Coal may be burned immediately after extraction but is most often crushed, sized, and cleaned to have many of its impurities (i.e., pyritic sulfur) removed before it is combusted. Further, coal can be either burned directly or converted into clean gaseous and liquid forms. Similarly, MSW can be used as "run-of-the-mine ore" but also may be prepared before combustion as a solid fuel. Further, it can be converted into gases or liquids before combustion. Incineration systems burn "run-of-mine" fuel, mechanical systems prepare and clean "urban ore" for direct combustion, and chemical conversion systems turn waste into gaseous and liquid fuels.

III. ENERGY RECOVERY SYSTEMS FOR URBAN WASTE

Energy recovery systems can be classified as mass burning or incineration, mechanical fuel cleaning and preparation systems, and conversion or pyrolysis systems. All three approaches are being utilized by many communities. Saugus, Massachusetts, for example, has a 1200-ton/day incinerator which provides steam to a nearby General Electric plant. Milwaukee and Chicago are among the many cities installing mechanical fuel preparation plants. South Charleston, West Virginia, and Baltimore, Maryland, have commercial sized plants converting urban waste into gaseous fuels by pyrolysis; Baltimore's system immediately burns the gas for steam production. In examining these available technologies, the discussion will utilize research systems where applicable, plus examples of commercial installations.

A. Run-of-Mine Combustion Systems

Run-of-mine MSW incineration has been used for many years, although energy recovery is relatively new in the United States. Incineration without heat recovery is practiced in many cities; some 30 million tons/year of municipal waste is disposed of in this manner. Several cities now practice energy recovery through incineration. For purposes of this discussion, the Saugus incinerator is utilized as the example [6].

Wheelabrator–Frye and the M. DeMatteo Construction Company joined in partnership to form the Refuse Energy Systems Company (RESCO) which operates the Saugus incinerator employing the Von Roll technology. Refuse is received in a 6700-ton-capacity pit. A traveling crane above the pit serves three functions: (1) mixing the waste to achieve some degree of homogeneity in the waste; (2) removing large, bulky objects and conveying them to a shredder; and (3) conveying the fuel to the incinerator. The crane also takes the shredded bulky objects back to the main waste stream after they are reduced to 12-in.-sized particles. When the waste is conveyed to one of two furnaces, it is burned at 1000 to 1800°F. The refuse is placed on the first of the three moving grates which are separated by steps. The movement of MSW along the grates, plus the tumbling between them, ensures complete burning of the organics. Combustion air flows from the refuse pit to the areas underneath and over the burning refuse, supplying sufficient oxygen for complete burnout, while controlling odors.

Von Roll incinerators produce steam in two stages. Refuse burning generates gases which heat the water walls of the boiler; the gases then pass through a convection section and come into direct contact with pendant boiler tubes for more efficient and complete energy recovery. Through this mechanism, combusting 50 tons of MSW produces 300,000 lb of steam with a temperature of 800°F and a pressure of 625 psi.

When cooled, the spent combustion gases pass through Wheelabrator–Lurgi electrostatic precipitators which reduce particulate emissions to 0.025 gr/scf (the Boston requirement, one of the most stringent in the nation, is 0.05 gr/scf). Since solid waste is an inherently low sulfur fuel, SO_2 removal is no problem. By burning the wastes at a temperature below 2000°F, the chlorides contained in PVC and other wastes do not decompose and re-form as hydrochloric acid, but remain as chloride salts. The bottom ash produced by the system is quenched in water and passed under a magnetic separation unit for ferrous metal recovery. The remaining ash is landfilled.

The Saugus plant, then, recovers both energy and ferrous metals while disposing of municipal wastes. Oil-fired packaged boilers exist, on a standby basis, to guarantee General Electric its supply of steam. Although the plant does not pollute the air or water, it requires landfill space to handle the 215 tons of residues produced daily, after the steel has been recovered.

Incineration of MSW suffers the same problems as combustion of run-of-mine coal. It releases large volumes of airborne particulates which must be captured by precipitators. Further, it produces similarly large amounts of bottom ash which must be disposed of. These problems created a need for coal cleaning technology, and they precipitated a move to mechanical beneficiation systems for MSW processing.

Despite its drawbacks, incineration has become a popular approach to energy recovery. In addition to the Saugus, Massachusetts, approach of providing process heat for manufacturing, incinerators have been connected to cooling systems. The Nashville Thermal Transfer Corporation 720-ton/day system in Nashville, Tennessee, supplies downtown customers with both steam for heating (in winter) and chilled water for cooling. A further process innovation has been proposed for northern New Jersey. There, Wheelabrator–Frye has designed an incinerator which is connected to an electricity generating power station, as depicted in Fig. 1.

B. Mechanical Beneficiation Systems

In the late 1960s and early 1970s, interest in producing cleaner solid fuels from waste emerged. The Union Electric Company developed, with EPA aid, a 325-ton/day simple prototype which has been frequently copied. It consists of a shredder, air classifier, and magnetic separation system. The U.S. Bureau of Mines, at its College Park, Maryland research facility, advanced this approach by applying conventional ore dressing systems to the problem of maximizing energy and materials recovery. Recognizing that

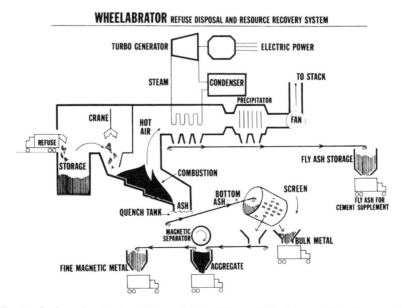

Fig. 1. Incineration for electricity production. (Source: Wheelabrator–Frye Company.)

low grade ores require multiple-stage processing, they liberated products by capitalizing upon differences in density, conductivity, magnetic susceptibility, and surface chemistry.

In the USBM dry mechanical process, waste enters the system through a course shredder which reduces particles to a −6 in. size. Waste then passes under a light air classifier where some of the paper and film plastics leave the product stream. From there the ore undergoes magnetic separation followed by primary air classification. In the primary air separator heavy iron objects which escaped magnetic separation fall out while light paper and plastics which escaped recovery in the first classifier are pulled off. A middling fraction of heavy organics, glass, stones, ceramics, and aluminum is also isolated. That middling fraction proceeds to a trommel where most of the glass, stones, ceramics, and heavy organics are removed. Through mineral jigging, the organics are removed for dewatering. Froth flotation sorts the glass from the stones and ceramics. The oversized material passing through the trommel is shredded to a −1 in. size, again air classified in an aspirator for final recovery of light paper and plastics, and the aluminum is separated from heavy organics by a combination of hysteresis and electrostatic separation. Figure 2 presents the USBM process flowsheet. Figure 3 depicts the primary separation of heavy iron, the middlings, and some of the fuel materials.

The fuel fraction from this process consists of the organics produced in the three air classifiers plus the heavy organics recovered in the jigging and aluminum recovery subsystems. Table III presents its characteristics.

Numerous vendors and engineers have produced a variety of mechanical systems. For purposes of this chapter, the American Can Americology system for Milwaukee is discussed. Its simplicity suggests the influence of the St. Louis prototype, yet several specific components reflect USBM research.

In the Americology system shown in Fig. 4, bundled newspapers and corrugated cardboard are handpicked from the waste stream before processing. Some 20% of the paper products can be removed in this manner, and returned to the paper industry. The mass of refuse then goes to a Williams hammermill for shredding to a 5-in. particle size. From there the refuse enters a zigzag air classifier—a contained vertical airstream moving through a series of diagonally placed baffles which create localized turbulence zones. Paper and light plastics move with the air through these zones and are collected at the top of the air classifier as fuel. The turbulence zones remove many glass fines produced by shredding which would otherwise contaminate the fuel fraction. Heavy particles drop to the bottom of the air classifier and are removed by a conveyor. These "heavies" then pass under a magnetic separation unit for ferrous metal recovery. The remainder

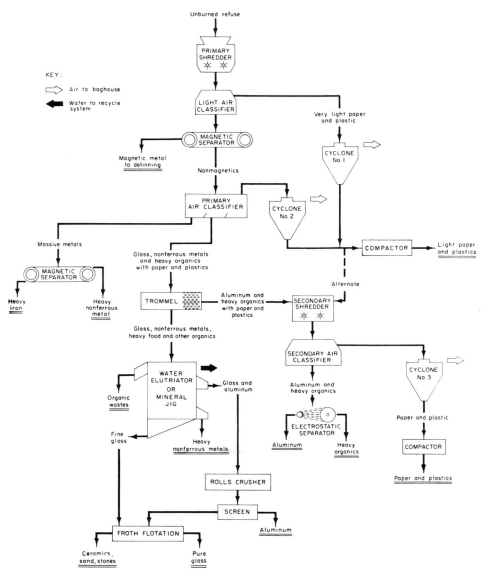

Fig. 2.   U.S. Bureau of Mines process flowsheet.

passes to a vibrating screen which removes the glass and stones. An "aluminum magnet" which works on the eddy current principle removes the light aluminum particles. Heavy organics remain for landfilling.

This system produces 84 tons/day of recovered paper fiber, 720 tons/day of supplementary fuel, 84 tons/day of ferrous scrap, 60 tons/day of glass

TABLE III

Fuel Evaluation of Refuse[a]

38	Heating value (BTU/lb)		Moisture (%)		Sulfur in total as-recovered combustibles (%)	Chlorine in total as-recovered combustibles (%)	Ash in total as-recovered combustibles (%)
	Total refuse, as-received	Total combustibles, as-recovered	Total refuse as-received	Total combustibles as-recovered			
High	5260	7430	32.0	24.6	0.30	0.70	15.2
Low	4260	6020	12.6	15.2	0.10	0.20	5.8
Average	4850	6630	20.8	20.7	0.20	0.50	10.3

[a] Source: U.S. Bureau of Mines.

Fig. 3. The primary separation of heavy iron, the middlings, and some of the fuel materials. (Source: U.S. Bureau of Mines.)

aggregate, 6 tons/day of aluminum, and over 240 tons/day of residues including the heavy organics. The relatively uncontaminated fiber may be returned to the paper industry. Wisconsin Electric Power Company utilizes the fuel fraction at their coal-fired Oak Creek generating station. The ferrous scrap is detinned to produce number one steel scrap and tin stannous. The aluminum is shipped to a secondary smelter. The residue is landfilled.

The fuels from this approach still contain relatively high percentages of moisture and ash, thereby necessitating combustion in coal-fired furnaces. Further, they are not homogeneous with respect to energy content; thus, best results are obtained by blending the refuse-derived fuel (RDF) with coal on a 10–15% RDF—90–85% coal basis.

A host of dry mechanical systems are being installed and operated around the country. Two 2000-ton/day units have been committed, one in Monroe County, New York, and one in Chicago, Illinois. Ames, Iowa has built and operates a 200-ton/day unit which utilizes the multistage process-

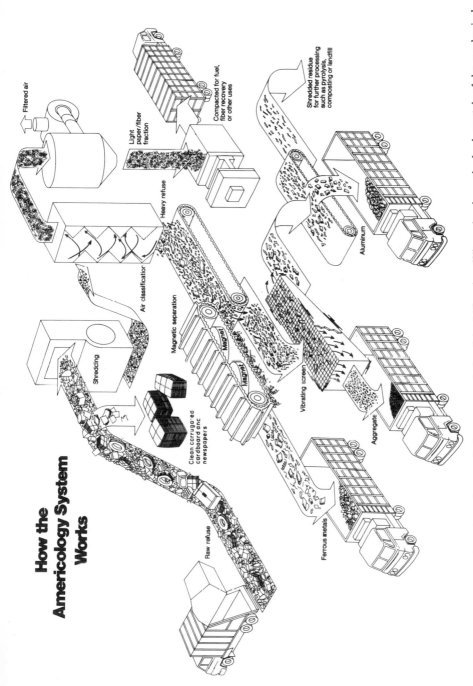

Fig. 4. This simplified schematic of the Americology system, designed for Milwaukee, Wisconsin, shows the fundamentals of dry mechanical processing technologies for resource recovery. (Source: The American Can Company.)

ing systems which are characteristic of the USBM technology. The Balti-more County, Maryland, dry mechanical system is almost a direct scale-up of the USBM technology. Other mechanical systems have been proposed, designed, and committed throughout the country.

The Black Clawson Company provides a novel approach to mechanical cleaning of urban waste fuels by using water rather than air as the carrying medium for the refuse during processing. In this approach, waste enters the system by being placed in a hydropulper, an oversized, highly powerful household type of blender. The ratio of waste to water is 3 to 97 by volume. After the hydropulper homogenizes the feed and reduces the particle size, the slurry leaves the bottom of the blender and enters a liquid cyclone which separates the organic and inorganic fractions on the basis of particle density. The organics, now homogenized, move to a dewatering system consisting of cones, presses, and screens. The result of this process is a fuel which is 50% water. Figure 5 illustrates the similarity of machinery between this fuel production system and papermaking. This fuel can be burned either

Fig. 5. This interior photograph of the Franklin, Ohio plant built by Black Clawson Fibre-claim demonstrates the similarity between wet pulping and the normal processes of papermaking. (Source: Black Clawson Company.)

as supplement to coal or separately, preferably in a bark-burning boiler. A 2000-ton/day Black Clawson system is being built in Hempstead, Long Island, which will produce steam for electricity generation.

C. Energy Conversion Systems

Like coal, energy-based fuels present their users with air pollution and residue disposal problems. Thus, numerous organizations developed systems for converting the impurity-laden solid fuels into clean, convenient gases and liquids. Research in chemical conversion began at the Pittsburgh Energy Research Center (PERC) in the late 1960s. There, the USBM–AGA retort for testing coal carbonization was adapted to waste pyrolysis. PERC demonstrated that destructive distillation could be used to produce clean gases and liquids from MSW efficiently.

Several firms brought commercial pyrolysis systems to the waste disposal field including Union Carbide Corporation, Carborundum Corporation, and Monsanto Chemical Company. Others, such as Occidental Petroleum and Atomics International, are in various stages of commercializing their systems, as suceeding chapters will discuss. Since the Union Carbide Purox process is the most advanced waste gasification system, it is discussed here.

Union Carbide Corporation developed the Purox pyrolysis system at its Tarrytown, New York research facility, and successfully operated a 5-ton/ day reactor there. Based on that success, it built the 200-ton/day module of a commercial plant in South Charleston, West Virginia, shown in Fig. 6. This test facility has been run on both as-received refuse and on shredded, magnetically sorted MSW. It has the capability for handling materials including sewage sludge dried to 20% solids, 80% moisture [7].

The Purox system is a vertical shaft furnace, not unlike a blast furnace. Wastes are charged into the reactor vessel at the top, and enter the first of three operating zones, the drying zone. By gravity they then move into the reaction zone where pyrolysis occurs. The gaseous and liquid products of this pyrolysis leave the reaction zone and move upward, while inorganics and char continue downward into the combustion zone. At the reactor hearth, the char is combusted with pure oxygen, fed through the tuyeres of the reactor, at temperatures of 3000°F. Figure 7 offers a schematic of this reactor and the various zones.

The abundance of char falling from the reaction zone creates a reducing atmosphere at the hearth, illustrated by the general equation $C + \frac{1}{2}O_2 \rightarrow CO$. The CO produced during combustion is released from the hearth of the reactor into the pyrolysis zone, moving countercurrent to the feed material to ensure maximum rates of heat transfer and gasification. Pyrolysis occurs in the reducing atmosphere created by the CO at temperatures ranging from

1200 to 1700°F. The rapid feed rates combined with the reducing atmosphere and the elevated temperatures ensure a minimum of oxidation and a maximum of gas production (with a minimum production of oil mists). Once the pyrolysis gas is produced, it moves upward through the feed material in the drying zone to the off-take pipe. This countercurrent flow of pyrolysis gases again ensures maximum contact with the incoming waste for drying purposes. The gases are produced at elevated temperatures, but leave the reactor at 200°F, as their sensible heat is used to drive moisture from the incoming feed material.

Once the gas has left the reactor it passes through a wet scrubber and precipitator cleaning system where oil mist is removed and returned to the reactor for cracking into additional gas, and fly ash is returned to the reactor where it is destroyed. After cleaning, the dry gas is sold to the customer. It has an average heating value of approximately 350 BTU/ft³, is free of potential pollutants, and performs in a manner similar to natural gas. Table IV presents its combustion characteristics compared to natural gas and propane.

Fig. 6. 200-ton/day module of a commercial plant in South Charleston, West Virginia. (Source: Union Carbide Corporation.)

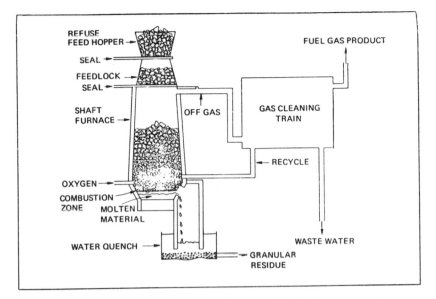

Fig. 7.   Schematic of the Purox process. (Source: Union Carbide Corporation.)

Two residue streams are produced by the Purox system: a molten slag or quenched frit stream which exits from the bottom of the reactor, and a wastewater stream resulting from the gas cleaning process train. The frit consists of the ash, metals, and glass contained in the original feed material. It passes through the combustion zone, where it is heated to a molten state and sterilized, and then quenched in a water bath. It can be either landfilled

TABLE IV

Comparison of Fuel Gas, Methane, and Propane[a]

	Purox fuel gas	Methane	Propane
Lower heating value (BTU/scf)[b]	350	910	2312
Compression power (kWh/mm BTU)[c]	5.7	1.8	0.6
Combustion air requirements (scf/mm BTU)[d]	8300	10,500	10,300
Volume of combustion products (scf/mm BTU)	10,500	11,600	11,200
Heat release/volume of combustion products (BTU/scf)	95	86	89

[a] Source: Union Carbide Corporation.
[b] Standard cubic foot dry, as measured at 60°F and 1 atm pressure. Heating value calculated at 18°C.
[c] Gas compressed to 35 psig from 1 atm, 100°F, with 75% efficiency.
[d] Based on the air needed to convert the fuel to CO_2 and H_2O—no excess air.

or used as fill material. The ash contains only 2% of the volume of the incoming refuse. The wastewater stream resulting from the gas cleaning train contains soluble organics which result in a chemical oxygen demand loading of 60,000 to 70,000 mg/liter. These are pretreated on-site before being released to the municipal sewage treatment system.

IV. TECHNOLOGY EVALUATION

Many cities cannot wait for second and third generation waste-to-energy systems. They must live in the here and now. Because these communities and regions face serious problems, they must evaluate currently available approaches. This evaluation must concern itself with which system, or systems, produces the highest quality and quantity of BTU for export to consuming industries. Energy product marketing, more than any other single factor, determines the success of an installation producing fuel from waste. A second consideration is the potential for pollution resulting from the production of fuels. This, too, must be considered.

A. Energy Production Evaluation

The use of waste as a fuel, or waste-based fuels, is inherently less efficient than the utilization of coal, oil, or natural gas. As Dr. Anderson demonstrated in Chapter I, the initial feed is far lower in quality. It is high in both moisture and ash, which are efficiency-inhibiting ingredients. To maximize energy production and compensate for the problems of moisure and ash, each system vendor has created an approach to allocation of efficiency losses. This allocation of losses, more than any other single factor, determines the price of the fuel.

The allocation of losses offers one approach to energy efficiency. For purposes of discussion, this paper calculates efficiency as

incoming energy in MSW (10,000,000 BTU)
   − mechanical losses caused by system
   − chemical conversion losses caused by system
   − energy required to run system
= export energy
   − efficiency losses in combustion of fuel
= net usable energy

$$\frac{\text{net usable energy}}{\text{incoming energy in MSW}} = \text{energy efficiency}$$

Those systems which mass burn, or incinerate, run-of-mine urban ore, absorb all of the inefficiencies themselves. They produce steam which can be used for space or process heat—directly—with no further combustion or treatment. Incinerator operations, on average, will produce about 5.2 to 5.8 million BTU of useful energy for every 10 million BTU contained in urban waste.

Mechanical systems absorb some of the energy inefficiency themselves. At the St. Louis prototype, for example, it was estimated that between 10 and 25% of the incoming BTU were discarded in the heavy fraction. The Milwaukee plant of Americology also discards the heavy fraction. Even with the beneficiation which the mechanically prepared fuel undergoes, the product is still high in moisture and ash content and heterogeneous in nature. Thus it can only be fired as a supplemental fuel in utility boilers. Further, these utility boilers must absorb energy efficiency losses. At Ames, Iowa, the efficiency loss due to the addition of this fuel is estimated to be 1.6% when the boiler is fired with 10% refuse and 90% coal. The losses are estimated at 2.2% when the mixture contains 20% refuse-based fuel. The estimates made by Rochester Gas and Electric for their Russell Station when utilizing fuel from the Monroe County recycling system are 1.46 and 2.58% when waste is 10 and 20% of the fuel feed [8]. Within this context it is important to note that a 3.3% efficiency loss will create an increase in fuel consumption of 10%.

Dry mechanical systems, on the average, produce a fuel which releases about 60 to 65% of the energy contained in the incoming waste material. This average considers the energy losses during preparation and combustion of the fuel. Wet pulped fuels, because they contain such a large quantity of water, are even less efficient. Of the energy entering a wet mechanical system, only 50 to 55% is released in useful form from combustion of the wet pulped fuel [9, 6, pp. 237–239].

Pyrolysis systems present a wide variety of energy efficiencies, depending on the particular process configuration and the product being maximized. The Union Carbide approach offers an efficiency range of 60 to 65%— equal to the energy production capability of the dry mechanical systems. It suffers an 8% loss to run the plant and equipment, a 17% loss in the chemical conversion process, and a gross loss of 10 to 15% associated with fuel combustion plus oxygen supply [9, p. 420].

B. Pollution Control Considerations

Pollution control is the second fuel quality concern. The effluents from a power plant or industrial boiler must be controlled, and the fuels that produce increased quantities of potential pollutants require more extensive application of pollution control devices. This use of pollution control devices increases capital expenditures associated with the utilization of waste-based fuels and, as a consequence, decreases economic value of those fuels.

Steam from an incinerator contains no inherent pollution, since the incinerator installation absorbs the entire responsibility for combustion of the refuse or the fuel derived from refuse. All other fuels, however, do contain varying quantities of potentially harmful constituents. Table V compares the chemical constituents of fuel from dry mechanical systems, wet mechanical systems, and the oxygen-fed chemical conversion system (10,9, p. 419].

From the chemical composition of these fuels certain generalizations can be made. These generalizations are supported by the experience of utilities and recycling systems operating in the waste-to-fuel industry.

Dry mechanical systems produce a fuel which not only has the limitation of being a supplementary fuel, but also the limitation of presenting a serious fly ash and bottom ash problem. As such it can only be burned in coal-fired stations with the capability for handling ash material. Even when burned in such a facility, the fly ash presents serious problems. At the Union Electric Company, for example, electrostatic precipitator efficiency declined from 97 to 93.5% when solid waste rose from zero to 30% of the fuel feed. Initial

TABLE V

Chemical Composition of Refuse-Derived Fuels

Fuel constituent	Percentage content		
	Dry mechanical (Union Electric)	Wet mechanical (Black Clawson)	Chemical (Purox)
H_2O	26.04	50	0
C	27.23	23.26	—
H	3.85	3.3	24 (as H_2)
CO	—	—	40
CO_2	—	—	25
CH_4	—	—	5.6
C_2H_x	—	—	2.4
Other hydrocarbons	—	—	2
H_2S	—	—	0.05
N	0.28	0.33	—
Cl	0.2	0.72	0
O	21.49	17.26	0
Ash	20.63	5.6	0
S	0.26	0.09	—

tests by that company indicated that firing the boiler with solid waste increased inlet dust loadings to the precipitator to 2.29 gr/scf—well within EPA regulations for existing power sources but above new source standards of 0.1 lb/mm BTU. With the cocombustion of coal and solid waste, Union Electric experiences 5.4 lb/mm BTU, whereas experience with 100% coal combustion is 4.9 lb/mm BTU [9, p. 422].

Wet pulped fuel offers fewer problems from an air pollution point of view. Like the dry mechanical fuel product, wet pulped fuel is low in sulfur. Its ash content is also relatively low. Still, either electrostatic precipitation or scrubbing equipment is required to remove the ash from the combustion gases before they enter the environment. Because Purox fuel is cleaned as it leaves the reactor, it produces virtually no air pollution. Combustion of Purox gas results in 0.008 grain of dry particulate per cubic foot (adjust to 12% CO_2), a rate below the EPA standard by a factor of 10 [11].

These generalized technology performance comparisons suggest that the degree to which waste is treated positively influences both its combustion efficiency and its potential for causing pollution. Solid fuels present the user with more combustion and pollution problems while decreasing the useful energy available to that product's customers. Pyrolysis fuels, in that regard, may offer superior performance.

V. TECHNOLOGY IMPLEMENTATION ANALYSIS

Within the space of a decade, inventors' dreams have become engineers' designs, engineers' designs have become city planners' hopes, and city planners' hopes have become reality. The decade 1965–1975 witnessed a surge of activity in the production of energy from solid waste.

At the same time, however, citizens' hopes often outstrip ongoing activities. In their perception of the enormous volumes of waste generated, and the depletion of available land for the safe and convenient discarding of those wastes, many citizens and public officials alike decry what they believe to be the slow rate at which resource recovery technology is being implemented on a commercial scale. This dichotomy of views raises serious questions regarding how fast energy-from-waste technology is being implemented, and if the pace of technology adoption is adequate.

Since the inception of this waste disposal approach, 14 plants have been built. These systems have the capacity to process 8515 tons of municipal waste per day—or 2,650,000 tons/year. These systems can deliver to the U.S. economy 2.0×10^{13} BTU/year. Currently, another nine systems are being built, having a capability for processing another 9100 tons/day of MSW—or 2,840,000 tons/year. When these plants go on-line, they will produce another 2.2×10^{13} BTU of energy for use in this country. By 1979 then, the recycling industry will be delivering 4.2×10^{13} BTU of energy. Table VI presents a list of recycling plants now operating, plus those under construction.

The energy delivery capability figures are small but significant for an economy searching for every available source of energy. More to the point, in the first 10 years of the energy-from-waste technology, this new industry captured 2.1% of the municipal waste disposal market. Within 15 years from the inception of this industry, based on present construction projects plus on-line plants, that market share will increase to at least 4.4%. Since those percentages are based on the total MSW generated, rather than that calculated to be economically available for recycling, the strength of this movement can be seen.

In addition to the plants now on-line or under construction, numerous others have passed through the "talking stage" and are being pursued vigorously by communities, regions, and private industry. Table VII presents a list of eight such projects. Their combined design capacity is 16,000 tons/day, or 4,576,000 tons/year (3.5×10^{13} BTU/year of energy production). Assuming that these systems, or ones equal in size, come on-stream by 1985, then the energy from waste industry will witness more than another doubling of capacity in the five years between 1980 and 1985. Such an assumption is conservative for projection purposes, particularly

TABLE VI

Waste-to-Energy Plants under Construction or Operational

Operational plants	Size (tons/day)	Plants being built	Size (tons/day)
Ames, Iowa	200	San Diego, California	200
So. Charleston, West Virginia	200	Hempstead, New York	2000
St. Louis, Missouri	325	Akron, Ohio	1000
Baltimore, Maryland	1000	Chicago, Illinois	1000
Baltimore City, Maryland	1200	Bridgeport, Connecticut	1800
Chicago, Illinois	1600	Milwaukee, Wisconsin	1000
Nashville, Tennessee	720	Monroe City, New York	2000
Harrisburg, Pennsylvania	720	Pompano Beach, Florida	100
Saugus, Massachusetts	1200		
Norfolk, Virginia	360		
Braintree, Massachusetts	240		
Ft. Wayne, Indiana	300		
Franklin, Ohio	150		
Bridgewater, Massachusetts	300		
	8515		9100

TABLE VII

Probable additional plants	Tons/day
St. Louis, Missouri	8000
Seattle, Washington	1500
New Britain, Connecticut	1800
Philadelphia, Pennsylvania	2000
Albany, New York	600
Hoosatonie Valley, Connecticut	1500
Madison, Wisconsin	200
Mt. Vernon, New York	400
	16,000
	4.0%

considering the fact that over half of today's landfill space will be exhausted by that date.

Absolute data provide one measure of growth in recycling. Another measure relates the growth of waste recycling to that of nuclear generation of electricity. Both new technologies stemmed, in large part, from governmental research efforts. Both addressed "public service" problems. Both fit into industries with monopolistic positions in their service area. Both substitute capital intensive solutions for more operating-cost-oriented solutions. Like most analogies, this one is limited. It is suggested only as a crude relative measure of new technology adoption.

The Atomic Energy Act permitting the use of nuclear power for electricity generating was passed in 1954, three years later the first plant went on-stream in Shippingport, Pennsylvania. In 1964, nuclear power accounted

TABLE VIII

Year	Municipal waste		Nuclear power (no. of plants)	Market share (electricity generation)
	No. of plants	Market share (waste disposal)		
0[a]	0	0	0	0
+5	1	—	1	—
+10	14	2.1	5	0.3
+15	22[b]	4.4	12	1.0

[a] Year for nuclear power is assumed to be 1954. For solid waste recycling it is assumed to be 1965, the year that the Solid Waste Disposal Act was passed.

[b] Estimates based on currently operating plants plus systems under construction or contract.

for 795 MW of electricity generating capacity. In 1969, nuclear power produced 1% of the electricity generated in the United States. This contrasts with the recycling industry which, in 10 years from its inception, captured 2.1% of its market and, within 15 years, will have captured 4.4% of the waste disposal market. Table VIII presents a comparison between these two technologies.

Similar comparisons can be made between the energy from waste technology and the introduction of the Basic Oxygen Process in the steel industry, or other major technical innovations. This crude technology adoption comparison suggests that the energy-from-waste industry is enjoying a rapid rate of commercial acceptance when compared to innovation in other industries.

VI. CONCLUSION

The development and adoption of technologies designed to convert municipal waste into useful fuels has proceeded rapidly. It has stemmed from a realization that waste creates environmental problems whereas a shortage of energy results in economic difficulty.

While these new technologies emerged, second and third generation systems began to travel the long road from the laboratory, through pilot and demonstration plant, toward commercialization. Production of liquid fuels is well along, as pointed out in Chapter V. Production of substitute natural gas by thermal processing, discussed in Chapter IV, is less advanced. Fluidized-bed combustion, an approach to industrial wastes discussed in Chapter XII, has been tried by Combustion Power Company with the 75-ton/day CPU-400 electrical generating pilot plant. It has not progressed beyond the large pilot stage.

While thermal processing and chemical conversion have advanced to commercialization, a new generic approach has begun to emerge. Biological processing appears to be a promising approach which will become commercial in the near future. Anaerobic digestion, which produces methane in all landfills, is particularly promising. In Los Angeles and San Francisco, old landfills have been drilled. Landfill gas is being extracted and cleaned up; essentially pure methane is the end product. In Pompano Beach, Florida, Waste Management, Inc. is building a 100-ton/day pilot plant under contract with the U.S. Energy Research and Development Administration. That plant will convert solid waste into substitute natural gas in a complex of digesting equipment. Less advanced is the enzymatic hydrolysis approach being pioneered by the U.S. Army Research Laboratory in Natick, Massachusetts. Biological conversion is also most applicable to feedlot wastes, demonstrated in Chapter VI.

Thus since 1965 and the recognition that municipal wastes threatened to inundate society psychologically if not psychically, three distinct approaches to the conversion of MSW into useful energy forms emerged. These approaches produce steam, solid fuels, and gaseous fuels. Continuing research and demonstration will provide systems to produce heavy oils, synthetic natural gas, and other energy products. But those communities "between the rock and the hard place" will adopt the technology that is now available. And they will do it at a rate which exceeds the adoption of other recently developed technical systems, for the solution to society's economic needs.

REFERENCES

1. Material needs and the environment today and tomorrow. The National Commission on Materials Policy, Cover letter, June 1973.
2. Third report to congress: resource recovery and waste reduction, p. 5. U.S. Environmental Protection Agency, 1975.
3. L. L. Anderson, Energy potential from organic wastes: a review of the quantities and sources. U.S. Bur. Mines, Information Circular 8549, 1972.
4. H. Bullis, Federal policy issues in the recycling of paper, pp. 42–3. Library of Congress Congressional Research Service, March 31, 1973.
5. R. C. Glaus et. al., The plastics industry in the year 2000. Stanford Research Institute, Stanford, California, April 1973.
6. W. K. MacAdam, Design and pollution control features of the Saugus, Massachusetts, steam generating refuse-energy plant. National Academy of Sciences, Mineral Resources and the Environment, Supplementary Rep., 1975.
7. J. E. Anderson, The oxygen refuse converter—a system for producing fuel gas, oil, molten metal and slag from refuse. Union Carbide Corporation.
8. Fuels from municipal refuse for utilities: technology assessment, pp. 3–23. Electric Power Research Institute, Bechtel Corp., March, 1975.
9. D. A. Tillman, Fuels from recycling systems, Environ. Sci. Technol. 9(5), 420 (1975).
10. J. D. Parkhurst, Rep. on status of technology in the recovery of resources from solid wastes. County Sanitation Districts of Los Angeles County, California, January 13, 1976.
11. D. M. Gilles, Union Carbide's purox process. National Academy of Sciences, Mineral Resources and the Environment Supplemental Rep., 1975.

Chapter III

PRODUCTION OF LOW-BTU GAS FROM WASTES, USING MOLTEN SALTS

S. J. Yosim and K. M. Barclay
ATOMICS INTERNATIONAL DIVISION
ROCKWELL INTERNATIONAL CORPORATION
CANOGA PARK, CALIFORNIA

I. INTRODUCTION

The disposal of municipal solid waste has become a critical problem from both the economic and the environmental points of view. The continued use of the traditional means of disposal (including open dumping, discharge into rivers, lakes, and occans, conventional incineration, and even sanitary land-

41

fill) is now being sharply questioned. Not only do these disposal methods have inherent air and water pollution problems, they are also wasteful of potentially valuable and recoverable energy and mineral resources. Typically, 50 to 60% of the constituents of the solid waste are combustible. In the United States, this represents 100 million tons of combustibles per year, equivalent to approximately 50 million tons of medium grade coal. This waste is available where the energy is required, in urban, industrial areas, without need for the additional energy normally expended to transport the fuel to its point of use. Thus, it is not surprising that obtaining fuel from wastes is receiving increasing attention nationwide. One type of fuel attainable from such wastes is low-BTU gas (100–150 BTU/scf) which can be used to generate process steam or, in the case of a large plant, to generate electricity.

Industrial wastes present equally serious problems, and equally promising opportunities. Wastes include synthetic rubber, waste plastic, x-ray film and other film materials, plus a host of other industrial wastes containing a high calorific content. Many of these wastes can yield high value metallic products as well as useful fuels. As an example, film pyrolysis may produce both silver and low-BTU gas if properly handled.

This chapter presents some experimental results which demonstrate the technical feasibility of applying the molten salt technology to the production of a low-BTU gas from wastes. The concept of molten salt gasification, with some of the advantages of this technology, is described first. This is followed by a description of the experimental apparatus used. Then some results of bench-scale gasification tests of selected wastes are given. Results include those from tests in which rubber tires, wood, and sucrose were gasified. Sucrose was selected as a convenient stand-in for wastes, which contain a considerable amount of oxygen (up to 50 wt %), because of its known composition and because it contains no ash.

The disposal of waste x-ray film is also discussed. Although low-BTU gas can be produced, a more important objective is to recover the silver from the film. Therefore, experiments to examine both objectives are described. The results of a pilot-scale gasification test with film are also discussed.

A discussion of the results of waste gasification, including a comparison of these results with those of coal and oil gasification using the same technology, concludes this chapter.

II. CONCEPT OF GASIFICATION OF WASTES WITH MOLTEN SALT TO PRODUCE LOW-BTU GAS

The concept of molten salt gasification of wastes is shown schematically in Fig. 1. Shredded combustible waste and air are continuously introduced

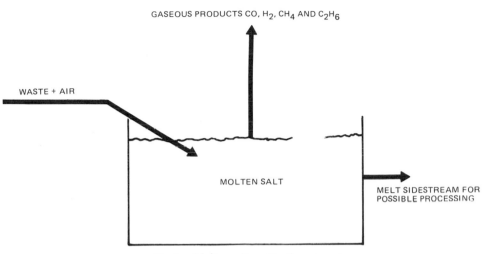

Fig. 1.   Molten salt gasification concept.

beneath the surface of a sodium carbonate-containing melt at about 1000°C. The waste is added in such a manner that any gas formed during combustion is forced to pass through the melt. Any acidic gases, such as HCl (produced from chlorinated organic compounds) and H_2S (from organic sulfur compounds), are neutralized and absorbed by the alkaline Na_2CO_3. The ash introduced with the combustible waste is also retained in the melt. Any char from the fixed carbon is completely consumed in the salt. The temperatures of gasification are too low to permit a significant amount of NO_x to be formed by fixation of the nitrogen in the air. Gasification of the waste is accomplished by using deficient air, i.e., less than the amount of air required to oxidize the waste completely to CO_2 and H_2O. Thus, the waste is partially oxidized and completely gasified in the molten salt furnace. The product gas flows to a conventional gas-fired boiler in which it is combusted with secondary air, producing steam.

As a possible option, a sidestream of sodium carbonate melt can be withdrawn continuously from the molten salt furnace, quenched, and processed in an aqueous regeneration system which removes the ash and inorganic combustion products retained in the melt and returns the regenerated sodium carbonate to the molten salt furnace. The ash must be removed to preserve the fluidity of the melt at an ash concentration of about 20 wt %. The inorganic combustion products must be removed at some point to prevent complete conversion of the melt to the salts, with an eventual loss of the acid pollutant-removal capability.

This concept is the basis of the molten salt coal gasification process which is currently being developed by Atomics International under an Energy Research and Development Administration (ERDA) contract.

In the application to silver recovery, the film is gasified in a Na_2CO_3 melt in the same manner as other combustible wastes. The silver from the film forms a liquid metal pool which is drained from the bottom of the combustor to form metal ingots with a purity exceeding 99.9%.

The advantages of molten salt gasification are as follows:

(1) Intimate contact of the hot melt, air, and waste provides for complete and immediate destruction of the waste.

(2) No HCl from plastics such as PVC or H_2S from sulfur-containing wastes is emitted.

(3) No tars or liquids are produced during gasification.

(4) Combustion products are sterile and odor-free.

III. EXPERIMENTAL SECTION

A. Materials Gasified

The waste x-ray film was analyzed and was found to contain, in weight percent: carbon, 53.2; hydrogen, 5.5; ash 2.4; and oxygen (by difference), 38.9. Direct analysis for silver showed that the film contained 2.3% silver. (Thus, the ash is essentially all silver.) The heating value on a dry basis was 8010 BTU/lb. The wood was pine sawdust with a moisture content of 2.8%. No chemical analyses were performed, but a typical composition for pine wood on a dry basis is, in weight percent, carbon, 51.8; hydrogen, 6.3; ash, 0.5; and oxygen, 41.3. The heating value is typically 9130 BTU/lb. The rubber was buffings from an automobile tire tread; no chemical analyses were performed. The nitropropane was practical grade obtained from Eastman. The sugar consisted of pure sucrose and had a heating value of 5940 BTU/lb. The coal was Kentucky No. 9 seam coal. The proximate analysis on a dry basis was, in weight percent, ash, 16.4; volatile matter, 37.6; fixed carbon, 46.0; and sulfur, 4.5. The ultimate analysis on a dry basis was carbon, 66.3; hydrogen, 4.6; nitrogen, 1.4; sulfur, 4.5; ash, 16.4; and oxygen (by difference), 6.9. The heating value on a dry basis was 12,000 BTU/lb. The fuel oil was a No. 6 oil, API gravity 18, carbon residue, 5%; ash, 0.007%; sulfur, 0.3%; and hydrogen, 13%. The heating value was 18,900 BTU/lb. The Na_2CO_3 for the salt bath was technical grade material obtained from Kerr McGee.

B. Bench-Scale Molten Salt Gasifier

A schematic of the bench-scale molten salt gasifier is shown in Fig. 2. Approximately 12 lb of molten salt is contained in a 6-in. i.d. and 30-in.-high alumina tube placed in a Type 321 stainless steel retainer vessel. This

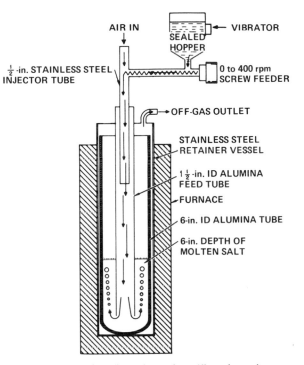

Fig. 2. Bench-scale molten salt gasifier schematic.

stainless steel vessel, in turn, is contained in an 8-in. i.d., four-heating-zone Marshall furnace. The four heating zones are each 8 in. in height, and the temperature of each zone is controlled by a silicon-controlled rectifier. Furnace and reactor temperatures are recorded by a 12-point Barber-Colman chart recorder.

Solids, pulverized when necessary in a No. 4 Wiley mill to < 1 mm in particle size, are metered into the ½-in.-o.d. central tube of the injector by a screw feeder. Rotation of the screw feeder is provided by a 0 to 400 rpm Eberback Corporation Con-Torque stirrer motor. In the injector, the solids are mixed with the air being used for gasification, and this solids–air mixture passes downward through the center tube of the injector and emerges into the 1½-in.-i.d. alumina feed tube. This alumina feed tube is adjusted so that its tip is ~ ½ in. above the bottom of the 6-in.-diameter alumina reactor tube. Thus, the solids–air mixture is forced to pass downward through the feed tube, outward at its bottom end, and then upward through ~6 in. of salt in the annulus between the 1½-in. and the 6-in. alumina tubes. In the case of liquids, a different feed system is used. The liquid is pumped with a laboratory pump and is sprayed into the alumina feed tube.

In order to prevent the melt temperature from rising when an excessive amount of heat is released to the melt, an air cooling system (not shown in Fig. 2) maintains a constant temperature. It consists of an eight-hole air distribution ring, mounted underneath the stainless steel ceramic tube retainer vessel. Air at rates up to 18 ft³/min can be passed upward between the outer surface of the retainer vessel and the furnace wall.

C. Analyses of Off-Gas from the Bench-Scale Gasifier

Samples of the exit gas, for analysis by gas chromatography, are taken with 1-ml gastight syringes downstream of the CO_2 analyzer. Two gas samples are taken at the same time. One sample is analyzed for carbon monoxide, oxygen, and nitrogen, using a molecular sieve $13\times$ column at room temperature. The other sample is analyzed for carbon dioxide, methane, ethane, ethylene, sulfur dioxide, and hydrogen sulfide, using a Poropak Q column at 130°C (266°F). The NO_x analyses are made with a Thermo Electron Corporation chemiluminescent NO_x analyzer. The carbon monoxide and carbon dioxide determinations are made with Olson-Horiba, Inc., units (Mexa-300 carbon monoxide and Mexa-200 carbon dioxide analyzers). All gas analyses data are reported on a dry basis.

D. Pilot-Scale Molten Salt Gasifier

A schematic of the molten salt pilot plant at the Santa Susana facility is shown in Fig. 3. Figure 4 is a photograph of the furnace.

The molten salt vessel, 10 ft high and 3 ft i.d., is made of Type 304 stainless steel, and is lined with 6-in.-thick refractory blocks. It contains 1 ton of salt, which corresponds to a depth of 3 ft, with no airflow through the bed. The vessel is preheated on start-up and kept hot on standby by a natural-gas-fired burner.

The salt loading is fed into the molten salt vessel through the carbonate feeder. The combustible materials to be processed are transferred directly from the hammermill, in which they are crushed to the required size, into a feed hopper provided with a variable-speed auger, and then introduced into the airstream for transport into the vessel.

Product or exhaust gases generated in the vessel exit through refractory-lined tubes in the vessel head to a refractory-lined mist separator. The separator traps entrained melt droplets on a baffle assembly. The gases are then ducted to the secondary combuster, which is 6-ft o.d. by 4-ft i.d. (The secondary combustor, which is between the furnace and the venturi scrubber, is not shown in Fig. 2.) A high energy venturi scrubber is used to remove any particulate matter before release to the atmosphere.

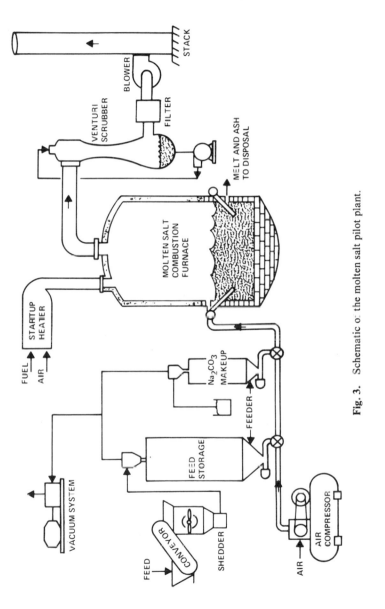

Fig. 3. Schematic of the molten salt pilot plant.

Fig. 4.   Pilot plant furnace.

IV.   GASIFICATION RESULTS

The gasification steady-state results for film, wood, rubber, and nitropro-pane are shown in Table I. The results for sucrose are shown in Table II, and those for conventional fuels (coal and fuel oil) are shown in Table III.

A.   Rubber

Two tests were conducted with buffings from a rubber tire. Since the tire contained organic sulfur which would form Na_2S in the melt, the Na_2CO_3 melt originally contained 6 wt % Na_2S to simulate steady-state conditions. Sodium sulfide was also added because it has been established that Na_2S is a catalyst in accelerating the gasification of char in molten Na_2CO_3. The gasification was accomplished with 33% theoretical air. The results in Table

TABLE I

Gasification of Wastes

Waste	Temperature (°C)	Air feed rate (scfm)	Fuel feed rate (lb/hr)	Percent theoretical air[a]	Composition of off-gas (vol %)					Higher heating value[b] (BTU/scf)
					CO_2	CO	H_2	CH_4	C_2	
Rubber	920	1.63	1.81	33	4.0	18.4	16.0	2.4	1.1	155
Wood	951	1.00	2.08	30	14.5	20.3	21.1	3.0	0.9	181
Nitropropane	1000	2.50	2.58	75	11	8	9	NM[c]	NM[c]	>55
Film	1015	4.50	5.34	51	16.5	12.0	11.7	2.6	0.2	107
Film	958	2.50	6.58	22	16.0	18.3	14.1	5.2	1.2	179

[a] Percentage of air required to oxidize material completely to CO_2 and H_2O.
[b] Calculated from composition of off-gas.
[c] Not measured.

TABLE II

Gasification of Sucrose in 100% Na_2CO_3[a]

Solid feed rate (lb/hr)	Percent theoretical air[b]	Composition of off-gas (vol %)						Higher heating value[c] (BTU/scf)
		N_2	CO_2	CO	H_2	CH_4	C_2	
12.2	18.4	34.3	13.5	26.9	18.9	4.3	1.9	226
6.5	34.8	46.2	15.0	18.3	17.0	2.5	0.8	154
4.3	52.2	57.0	19.1	12.3	9.7	1.4	0.3	91
3.1	72.1	69.8	20.0	5.5	3.1	0.7	0.1	37

[a] The melt temperature was 950 to 960°C; the air feed rate was 2.5 scfm.

[b] Percentage of air required to oxidize material completely to CO_2 and H_2O.

[c] Calculated from composition of off-gas.

I are an average of the two tests at identical conditions. A gas with a higher heating value (HHV) of 155 BTU/scf was obtained. The CO_2 content of this gas was considerably lower than that from the oxygen-containing wastes and is more similar to that obtained when coal or oil is gasified. No H_2S or other sulfur-containing gases (< 30 ppm) were detected in the off-gas.

B. Wood

Gasification of the pine wood was accomplished with 30% theoretical air in pure Na_2CO_3. Again, a gas with a high heat content (181 BTU/scf) was obtained. The H_2 content was somewhat higher and the CH_4 content somewhat lower than was obtained with the film at 22% theoretical air.

C. Sucrose

Sucrose was selected as a stand-in for oxygen-containing wastes. The gasification of sucrose in pure Na_2CO_3 was studied at four different stoichiometries. The melt temperature was about 950°C. The air feed rate (2.5 scfm) corresponded to an air superficial velocity of 1 ft/sec in the gasifier. As expected, the heating value of the gas increased as the percentage of theoretical air decreased. However, when the air/fuel ratio was too low, insufficient heat was released to the melt and auxiliary heating was required. This was the case with the test in which 18% theoretical air was used. The furnace was turned on during the test to maintain the melt temperature. However, at 35 and 52% theoretical air, sufficient heat was released to the melt so that the furnace could be turned off. At 72% air, excess heat was generated and had to be removed by the cooler. The relatively short experi-

TABLE III

Gasification of Coal and Fuel Oil

Fuel	Temperature (°C)	Air feed rate (scfm)	Fuel feed rate (lb/hr)	Percent theoretical air[a]	Composition of off-gas (vol %)					Higher heating value[b] (BTU/scf)
					CO_2	CO	H_2	CH_4	C_2	
Coal[c]	983	1.63	2.53	33	3.5	30.8	13.7	1.2	0.11	158
Coal	988	1.73	2.27	40	6.4	27.0	12.2	1.1	0.08	139
Coal	985	1.90	2.13	47	8.4	23.0	10.8	0.7	0.05	117
Coal	987	2.14	1.56	75	15.3	6.3	2.9	0.6	0.05	37
No. 6 fuel oil	985	1.63	1.98	27	10.7	11.0	17.3	3.3	1.8	157
No. 6 fuel oil and thiophene[d]	925	1.63	2.68	20	4.5	14.5	19.3	5.5	1.6	194

[a] Percentage of air required to oxidize material completely to CO_2 and H_2O.

[b] Calculated from composition of off-gas.

[c] The melt contained 15 wt % Na_2S in all cases except the first test with No. 6 fuel oil. In that test, the melt was pure Na_2CO_3.

[d] Enough thiophene was added to result in 5% sulfur in the oil.

mental runs (~1 hr) did not permit a definite conclusion as to what minimum percentage of theoretical air could be used in this system and still maintain melt temperature without auxiliary (furnace) heat. However, it appears that the salt could be maintained in a molten state at least at 35% of theoretical air, at which level a combustible gas with an HHV of 154 BTU/scf was being generated.

A test in which the air superficial velocity was reduced to ½ ft/sec yielded an off-gas virtually unchanged in composition from that when the velocity was 1 ft/sec. This suggests that residence time of air in the melt is not an important factor.

D. Nitropropane

Organic NO_2-containing compounds may well be present in small amounts in wastes which are being gasified for the production of fuel. This type of compound can emit a great deal of NO_x. Nitropropane was used in these tests to determine if it would be possible to minimize the NO_x emissions during gasification. A series of tests were run under varying conditions and with different additives. It was determined that under certain process conditions and with a specific additive, it was possible to combust nitropropane and produce a product gas containing only 6 ppm NO_x. It thus appears that the presence of organic NO_2-containing wastes will not contribute much NO_x under gasifying conditions. The composition of a typical off-gas from the tests is shown in Table I.

E. Film

Two series of tests were conducted with waste x-ray film. The first series was performed in the bench-scale gasifier; the second series was run in a pilot plant which is capable of gasifying about 200 lb/hr of coal and waste. Two tests were run in the first series. The purpose of the first test (51% theoretical air) was to show that pure elemental silver could indeed be recovered under gasification conditions attainable in the Atomics International pilot plant. A combustible gas with an HHV of 107 BTU/scf was obtained. In this bench-scale test, 15 lb of film were burned. After the test, the melt was cooled until the silver solidified (~960°C). A bright pellet of lustrous silver metal having a weight of 0.34 lb was recovered. This corresponds to 99% of the silver fed.

In the second test, the gasification was accomplished under more reducing conditions (22% theoretical air). This time a gas with a much higher heat content was obtained (179 BTU/scf). Elemental silver was recovered from this test also but the yield was not determined.

In the pilot plant test, 15,000 lb of waste x-ray film was burned at the rate of 200 to 230 lb/hr. The average air feed rate was 180 scfm. The air/film ratio corresponded to 50% theoretical air. Since the main purpose of this test was to recover silver, no attempt was made to analyze the off-gas. After the film had been gasified, the silver and salt were drained from the vessel and 354 lb of 99.97% pure metallic silver was recovered. A single 250-lb silver ingot is shown in Fig. 5.

F. Conventional Fuels

The results of gasification of coal and fuel oil are shown in Table III. The most extensive work was done with coal. The dependency of the heat

Fig. 5. Silver ingot produced from waste x-ray film.

content of the gas on the percentage of theoretical air can be seen in Table III. As expected, the heat content of the gas increases as the percentage of theoretical air decreases. No H_2S or other sulfur-containing gases (< 30 ppm) were found in the off-gas. This was found to be the case even when fuel oil contained substantial amounts of thiophene (corresponding to 5 wt % sulfur).

V.   CONCLUSION

A comparison of the heating values of the product gases obtained from all the gasification tests shown in Tables I–III is given in Fig. 6. As shown, the heating value of the product gas decreases in a smooth fashion as the percentage of theoretical air increases. It is also evident that the heating values of the gas produced by gasification of these materials are essentially independent of the composition of the fuel at a given fraction of theoretical air. This is illustrated in the case of coal (7 wt % oxygen) and sucrose (51 wt % oxygen) at about 35 and 70% theoretical air. This overall correlation exists in spite of the large difference in air superficial velocities, ranging from 0.25 ft/sec in the case of wood to 1.6 ft/sec in the case of film at 51%

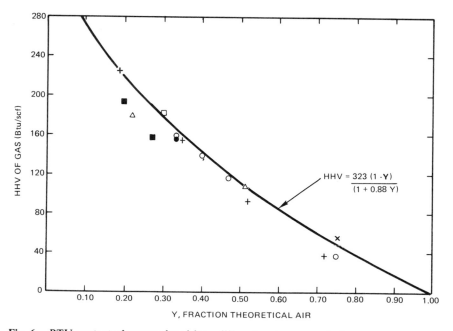

Fig. 6.   BTU content of gas produced from different carbonaceous fuels: +, sucrose; △, film; □, wood; ●, rubber; ○, coal; ■, fuel oil; ×, nitropropane.

theoretical air. Figure 6 also includes a curve which corresponds to the expression for the HHV:

$$HHV = 323\left(\frac{1 - Y}{1 + 0.88Y}\right)$$

where Y is the fraction of theoretical air. This expression is essentially that derived from considering only CO and H_2 as the products from the gasification reaction. A rigorous derivation has been performed, taking into consideration all the gaseous products formed. It was found that the neglected terms canceled each other to a significant degree.*

While for a given stoichiometry, the heat content of product gas per standard cubic foot is similar for the materials studied in this program, the material without oxygen will produce more total product gas per pound of material. Therefore, as would be expected, the total heat extracted from such material would be greater than from material containing oxygen.

The effect of temperature on the heat content of the product gas is not clear. There did not appear to be any correlation between the deviation of the data from the approximation curve and the temperatures of the tests.

It has been shown that, as expected, the BTU content of the off-gas increases as the percentage of theoretical air decreases. However, as was also shown (in the tests with sucrose), there is a practical upper limit on the gas heat content which can be obtained. Above this upper limit, there will not be enough heat released to the melt to sustain the operating temperature. This upper limit can be increased somewhat by using preheated process air and by decreasing the heat losses in the gasifier with improved insulation.

The maximum waste throughput is also governed by the maximum superficial velocity of the gas through the melt. In general, the maximum super-

---

* The rigorous derivation was accomplished by considering a general combustible waste containing carbon, hydrogen, and oxygen. Gasification of the waste with a fraction Y of theoretical air produced CO, H_2, CH_4, and C_2H_6, as well as CO_2 and H_2O. By carrying out a carbon, hydrogen, and oxygen balance and using the mole fractions X of CO_2, CH_4, and C_2H_6 as the independent variables, the following expression was obtained:

$$HHV = \left[\frac{323(1 - Y)}{(1 + 0.88Y)}\right] + \left[\frac{323(1 - Y)}{(1 + 0.88Y)} \left\{- (X_{CO_2} - 3X_{CH_4} - 6X_{C_2H_6})\right\}\right.$$
$$\left. + 323(4X_{CH_4} + 7X_{C_2H_6})\right] + [10.2X_{CH_4} + 17.7X_{C_2H_6}]$$

Of the three terms in brackets, the first two represent the contribution of CO and H_2 to the heating value, and the last term represents the contribution due to CH_4 and C_2H_6. The second and third terms, which are neglected in the approximation, in general have opposite signs, which results in cancellation of a significant portion of their absolute values.

ficial velocity of the inlet air has been set at 2 ft/sec. (This corresponds to a somewhat higher velocity of the product gas, depending on the fraction of total oxygen which is combined oxygen and on the composition of the off-gas.) Beyond this velocity of 2 ft/sec, entrainment of the melt becomes excessive. However, by operating at elevated pressures, the waste throughput can be significantly increased since at a given air superficial velocity, the waste throughput will be proportional to the pressure.

No sulfur-containing pollutants were observed in the off-gas when sulfur-containing material was gasified. Although chlorine-containing material was not gasified, in all tests in which this type of material was treated in Na_2CO_3 melts with excess air, there was no trace of HCl, even when the Na_2CO_3 content was as low as 5 wt %. It is expected that the same results would be obtained if deficient air were used.

The results described in this chapter show that the gasification of wastes in molten salts to produce a low-BTU gas is technically feasible. However, an engineering evaluation leading to the economics of molten salt gasification of the various wastes has not been done. It is anticipated that the costs for processing municipal wastes will be greater than those for conventional incineration. However, for special problem wastes such as hazardous wastes, the costs may be competitive.

Chapter IV

PIPELINE GAS FROM SOLID WASTES BY THE SYNGAS RECYCLING PROCESS

Herman F. Feldmann, G. W. Felton, H. Nack
BATTELLE-COLUMBUS LABORATORIES
COLUMBUS, OHIO

J. Adlerstein
SYNGAS RECYCLING CORPORATION
WEST TORONTO, ONTARIO, CANAD

I. INTRODUCTION

Currently, most cities in the United States are faced with a severe shortage of natural gas that is curtailing industrial operation and expansion, reducing the fuel supplies of institutions such as schools and hospitals, and forcing new home construction to utilize expensive electrical energy for heating and cooling which results not only in higher prices but also causes a very low resource utilization efficiency.

Unfortunately, it is apparent that the natural gas supply problem will not be confined to the United States. For example, it is anticipated that, even in energy-rich Canada, the shortage will be felt in eastern Canadian metropolitan areas as their growth and energy consumption exceed the available natural gas supply. In anticipation of this problem, Canada is already reducing exports of natural gas to the United States and seriously considering gasification of coal in the western provinces to reduce the impact of the natural gas shortage. Also, there is reason to believe that with rapid conversion to natural gas in England and on the continent because of the North Sea discoveries, Western Europe will also experience the same natural gas shortages now widespread in the United States.

Ironically, in the metropolitan areas, where the gas shortages are most severe, the disposal of an ever increasing volume of solid wastes from municipal, industrial, and sewage treatment operations is also a growing problem of major proportions.

Conversion of this solid waste into a synthetic natural gas (SNG) in an environmentally acceptable fashion would, therefore, assist in the solution of two very severe urban problems. Conversion of solid waste into a fuel gas of much lower BTU value than natural gas is possible with known processes and apparatus but technology for the production of synthetic natural gas from solid waste gas has not been available heretofore. The only process currently available for converting solid wastes into synthetic natural gas is by biological digestion. The two main problems with biological digestion are the following:

(1)   the long solid waste residence time, which requires extremely large vessel capacity if large volumes of waste are to be treated, and

(2)   disposal of the by-product sludge from the process.

Thus, the objective of the work reported here is to develop a process that allows solid wastes of municipal, industrial, or agricultural origin to be converted to a raw gas that can be upgraded to SNG using available gas purification and methanation technology.

Previous batch autoclave studies [1] established that solid wastes could be hydrogasified and that, depending on the hydrogen/solid waste feed ratio, methane and ethane yields would be substantial. These experiments established the feasibility of using solid wastes as a feedstock for hydrogasification. An accompanying very preliminary cost estimate indicates that for larger municipalities hydrogasification of solid wastes to pipeline gas should be very attractive economically.

Because these preliminary technical and economic evaluations were positive, an extensive process development effort was initiated for the following purposes:

(1) to develop the most attractive reactor systems for carrying out the conversion on a commercial scale,

(2) to generate continuous data with which to design a demonstration plant, and

(3) to perform additional technical and economic analyses based on the concepts and data arising from this process development effort.

The process resulting from this development effort offers the following advantages for the conversion of municipal wastes to SNG:

(1) No preseparation of metal and glass is required.

(2) Higher yields of methane are produced than are possible by alternate technology.

(3) Gas is produced at the proper pressure for purification.

(4) The recycle value of the metal and glass is preserved because no oxidation or sintering occurs.

(5) A relatively small volume of disposable residue is produced.

(6) The process has the flexibility to handle ranges in feedstock composition.

II. THE PROCESS CONCEPT

Before initiating the program, many potential reactor schemes were analyzed in order to establish what system or systems offered the most advantages for commercially converting solid wastes to a methane-rich product gas. The following are some of the criteria applied for the evaluation of these reactor system candidates:

(1) Allow the use of unseparated solid waste.

(2) Allow the production of a raw product gas comparable to that from a Lurgi gasifier so that similar gas purification and methanation systems can be employed.

(3) Allow the resource value of the metal and glass to be preserved.

(4) Be simple enough to ensure a high degree of reliability.

(5) Allow scale-up from the size used in these experimental studies to a demonstration-sized plant.

Without going into the details of these evaluation studies, the reactor scheme shown in Fig. 1 was conceived as the most promising.

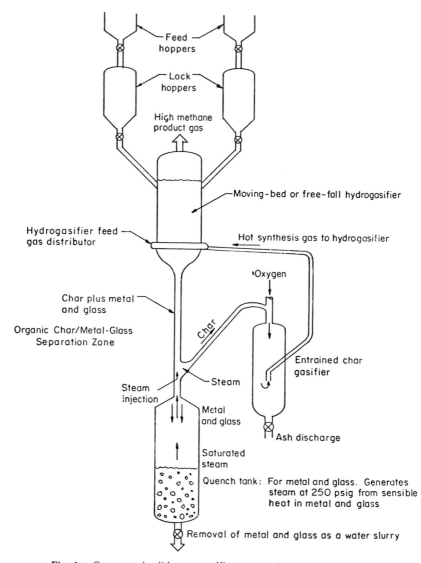

Fig. 1.   Conceptual solid waste gasifier system. Drawing is not to scale.

In this concept, solid waste is shredded and lockhoppered into pressurized hoppers. From the pressurized hoppers it is fed into a methane production reactor (MPR) where it encounters a countercurrent stream of hot hydrogen-containing synthesis gas. In the MPR, the solids can either fall through in the free-fall mode or move downward through the reactor as a moving bed.

After leaving the MPR, the solids fall through a stripping zone where steam entrains the very light (compared to the metal and glass) organic char. The char is blown into a gasifier into which is fed oxygen to gasify the char, and the hot synthesis gas is then fed directly to the MPR.

The metal and glass after falling through the stripping zone enter a quench pot and are discharged from the reactor system in a water slurry from which they are separated for recovery and recycle. Some of the more significant features in this system are as follows:

(1) Flexibility in the separate control of methane production and gasification reaction zones exists.

(2) Methane production requires considerably milder conditions than gasification.

(3) Optimum contacting schemes for methane production were conceived to be different from those for gasification.

(4) The system allows separation of the metal and glass from the organic fraction after the methane production zone. At this point, because of the devolatization of the waste, there is a great density difference and the metal and glass are degreased because of the high temperature and contact with the reducing gases.

(5) Physical separation of the methane production and gasification zones will ensure that no methane is burned by the oxygen or reformed by the steam fed to the gasifier.

The two contacting schemes selected for investigation in the methane production reactor were free fall and moving bed. Countercurrent gas-solids flow was also selected as being the most promising because of the more effective heat recovery from the raw product gases.

Because of the availability of data on the gasification of carbonaceous char, the key data needed for the design of the reactor system shown in Fig. 1 were for the MPR. Therefore, the experimental program focused on the devolatization and hydrogasification of the solid waste.

III. EXPERIMENTAL STUDIES

A. Objectives

The objectives of the experimental portion of the process development program were to

(1) determine whether and under what conditions an acceptable gaseous product can be produced,

(2) determine the yields and compositions of side products,

(3)   evaluate potentially attractive contacting schemes,

(4)   determine the critical factors affecting scale-up, and

(5)   generate the necessary data for a "base case" design.

B.   The Reactor System and Operational Procedure

As mentioned, the objective of the experimental portion of this program is to define the effect of operating parameters on the operation of a continuous hydrogasifier feeding a "typical" solid waste.

The solid waste used in these experiments was of a standard composition recommended by the Environmental Protection Agency for incineration studies that was modified somewhat to allow passage through the 2.8-in.-i.d. methane production reactor without bridging. Typically, the solid waste used in these experiments had the elemental balance shown in Table I.

The reactor used in these studies was a tubular reactor of 2.8 in. i.d. with a heated length of approximately 10 ft. The reactor was made from a length of reformer tube 2.8 in. i.d. with appropriate connections machined at both ends. The shredded solid waste was charged into a feed hopper and, after charging, the entire system was purged with inert gas and then pressurized with hydrogen to the desired run pressure. Heat was supplied to the reactor by means of an external resistance furnace. When the desired temperature was reached, hydrogen flow was initiated and the solids fed into the reactor from the feed hopper by means of a screw feeder. A sketch of the reactor system is shown in Fig. 2.

All operations were made with countercurrent contacting of the solid waste and feed gas. As explained, the reason for countercurrent operation with plug flow of solids was based on heat balance considerations. For

TABLE I

Composition of Standard Solid Waste Used[a]

Constituent	Weight percent
Carbon	35.2
Hydrogen	4.9
Nitrogen	0.2
Oxygen	35.5
Ash	1.3
Moisture	22.9
	100.0

[a] A mixture of approximately 25 wt % raw potatoes and 75 wt % paper.

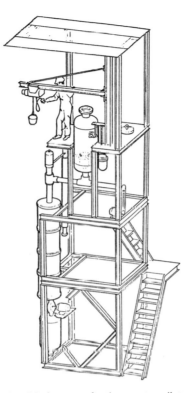

Fig. 2. Methane production reactor pilot unit.

example, such systems allow the greatest internal heat recovery from the raw product gas and minimize the effect of varying moisture content on temperatures in the hydrogasification zone of the reactor. Both free-fall and moving-bed solids–gas contacting schemes were evaluated.

The liquid products, which turned out to be mostly water, were condensed from the gases and collected after the experiment was completed. After passing through micropore filters, the gas volume was measured with a dry test meter and samples taken at approximately 5-min intervals for "on-line" gas chromatographic (GC) analysis. In addition, several spot gas samples were taken during the steady-state portion of the experiment for mass spectrographic analyses as a check on the on-line GC analyzer.

Both moving-bed and free-fall solids flow systems were evaluated. In both cases, the solids after undergoing hydrogasification were fed by means of a rotating feeder into a solids receiver which remained at essentially ambient temperature. After the experiment, the solid char resulting from hydrogasification was weighed and analyzed.

One of the problems in these experiments was the feeding of solids at a uniform rate. Because of bridging the solids feed rate would sometimes change during the course of an experiment. That this was occurring at times was obvious from observing the variation in product gas composition. Checks on total product recovery and analytical procedures were made by collecting the total material including all the gas inventoried in the reactor at the run end. These indicated that the data generated are reliable and within the accuracy of ordinary experimental error.

Because of the continuous metering and analysis of the product gas, it was felt that it would be most logical to use the average gas flow rate and analysis together with the ultimate analysis of the char and the char/feed waste ratio to calculate the effective solids feed rate over the steady-state portion of the experiment by a carbon balance. The results presented are based on this solids rate. Ordinarily there was very good agreement between the feed rate calculated on the basis of a carbon balance and that calculated by dividing the total solid waste fed by the time the screw feeder was turned on. Another potential source of error was the segregation of waste in the hopper with the denser potatoes tending to feed preferentially. However, this was corrected for by reporting results on a dry basis because, on a dry basis, the ultimate analysis of the various waste constituents is very similar.

Since the objective of this program was to provide the basis for the design of a demonstration or large pilot plant rather than a process variable study, experiments were made at conditions of greatest commercial potential.

One of the most critical parameters to specify was the system pressure. Initial experiments were made measuring the total rate of hydrogasification in a pressurized thermal gravimetric analysis unit (TGA) at pressures of 100, 300, and 500 psig. The results of these experiments are summarized in Fig. 3 where it can be seen that the total rate of gasification was not greatly influenced by the system pressure.

Thus, a pressure of 250 psig was selected for our pilot plant experiments because we feel this pressure offers a good compromise between capital investment, gas compression, and purification costs, and the ability to utilize existing lockhopper technology to pressurize the waste.

The critical question was, "At this operating pressure can a reasonable methane content product gas be produced and further can the production of tars be kept to acceptable levels?" As the results that follow will show, the answer to these questions is indeed yes and further that either moving-bed or free-fall contacting schemes may be employed.

In addition to the pressure, other system variables that are independently controllable in the pilot plant are temperature, H_2/waste feed ratio, and solids residence time.

As previously mentioned, the reactor pressure was selected to allow what

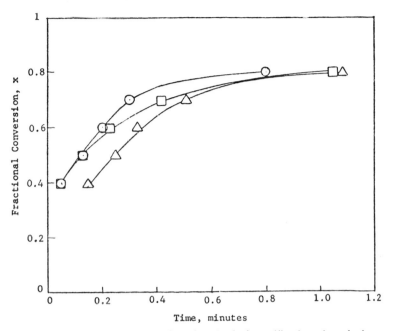

Fig. 3. Fractional conversion versus time data for hydrogasification of synthetic waste at 1000°F. Data points: O, 500 psig; □, 300 psig; △, 100 psig.

was felt to be the best balance between solids feeding, gas purification operating costs, gas compression, and total investment required for the plant. In addition, shakedown operations indicated that reactor wall temperatures of 1600°F were needed in free-fall operation to ensure that in free fall the shredded waste was heated to reaction temperature. For example, in trial runs at a reactor wall temperature of 1200°F only the outer surfaces of the solid waste particles appeared carbonized, with most of the particle interior never reaching the temperature at which carbonization was initiated. Thus, most of the experiments were conducted at reactor wall temperatures near 1600°F.

Results of the pilot plant experiments are summarized in Table II for both moving-bed and free-fall operations.

IV. SUMMARY OF EXPERIMENTAL RESULTS

A. Contacting Schemes and Hydrogen/Waste Feed Ratio

1. Effect on Carbon Conversion The mode of contacting, either free fall or moving bed, had the greatest influence on the results of these experi-

TABLE II   *Summary of Typical Methane Production Reactor Pilot Plant Operation*

	Free-fall experiments			Moving-bed experiments		
	Run 5	Run 6	Run 7	Run 9	Run 10	Run 11
Solid waste feed rate						
Dry basis (lb/hr)	7.39	10.50	7.04	8.69	6.70	6.12
As-fed basis (lb/hr)	10.40	13.46	9.76	10.89	8.91	9.24
Hydrogen feed rate (scf/hr)	180	98	100	100	50.0	44.2
Pressure (psig)	250	250	400	250	250	250
Reactor temperature (°F)	1600	1600	1600	1550 [a]	1600 (871°C)	1500 (816°C)
Approximate solid waste residence time (sec)	1	1	1		21 min	13 min
Carbonaceous char yield (lb/lb as-fed waste)	0.214	0.271	0.241	0.147	0.154	0.173
Oil yield: (%) (lb/lb waste as-fed)	<1	NM[b]	<1	0.5		
Water yield (lb/lb dry waste)	0.408	0.282	0.279	0.253	0.008	0.008
Product gas yield (scf/lb dry waste)	25.6	16.3	17.6	16.2	14.0	11.9
Product gas composition[c] (vol %, dry basis): H_2	68.5	61.1	58.6[d]	46.0[e]	32.42	33.8
CH_4	12.5	12.9	17.2	20.7	27.34	23.8
CO	15.0	17.4	17.1	23.4	24.20	24.7
CO_2	3.0	4.6	6.0	7.0	13.30	13.2
C_2H_6		1.5			1.50	1.9
C_2H_4		0.1				1.3
C_3H_8						0.2
C_6H_6		0.6			0.50	0.7
N_2	1.0	1.8	1.1	2.9	0.74	0.4
Total	100	100.0	100.0	100.0	100.00	100.0

[a] Could not be determined for this run.   [b] Not measured (<1%).

[c] Results for Runs 7 and 9 are for average steady-state product gas composition.

[d] Hydrocarbons higher than CH_4 were not analyzed for in Run 7.

[e] Standard cubic foot for Run 9 measured at 1 atm and 60°F (15.6°C).

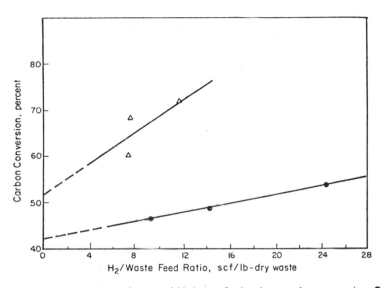

Fig. 4. Effect of contacting scheme and H_2/waste feed ratio on carbon conversion. ●, Free fall; △, moving bed.

ments. For example, the carbon version achieved in the MPR is shown in Fig. 4 for both free-fall and moving-bed contacting systems. As was mentioned, there was difficulty in controlling bed height in the moving-bed systems* and so the experiments were made at a variety of residence times that ranged from perhaps a minute or so in the very shallow bed in Run 9 to 20 min in Run 10. (The residence times reported were based on the amount of solids in the reactor at the conclusion of an experiment.) In addition, there were indications that the bed heights may have varied considerably while the experiment was in progress due to rather erratic discharge of the hydrogasified char. In spite of the periods in which there was erratic solids discharge, the gas rate and composition during the run remained steady as long as the solid waste feed rate into the reactor remained constant.

Based on the apparent independence of conversion and the examination of the char in the free-fall experiments which indicated that even with a 1600°F reactor wall temperature interior particle temperatures only reached the point of incipient charring, it appears that the most critical parameter influencing conversion is the heat transfer rate.

---

* Again this problem is one associated with the small scale of the equipment. For example, knowing where the bed level is at any instant of time is a problem in our small reactor. In a larger reactor nitrogen-purged large-diameter ΔP lines could be used but in the existing plant this would cause large amounts of product gas dilution. In addition, the solid waste displays uncanny ability to bridge small-diameter reactors.

Another important parameter influencing carbon conversion is the hydrogen/solid waste feed ratio. Increases in this parameter also result in increases in the carbon conversion. A plausible explanation for this behavior is that the hydrogen must be available to react with volatile intermediates which otherwise tend to crack to residual carbon. As the tabulated data indicate, the formation of tars is extremely low and if any variations in the production rate of these tars with charging hydrogen/solid waste feed ratios occur, they cannot be detected at this scale.

In an integrated plant, the MPR is combined with a gasification reactor which provides the hydrogen (as synthesis gas) and the volume of synthesis is proportional to the carbon in the char exiting the MPR. Thus, a balanced operating point may be calculated by selecting the kind of gasifier to generate the synthesis gas. For the purpose of calculating this balance point, we selected an entrained gasifier system because the low bulk density of the char from the MPR makes it easy to entrain and because of the reliability of entrained gasifiers.

Figure 5 contains the experimental curves presented in Fig. 4 together with a calculated curve showing the hydrogen in the feed gas that can be produced from the carbon in the char exiting the MPR. Data presented by VonFredersdorff and Elliott [2] for a Texaco entrained gasifier were used to calculate the hydrogen production curve. Under actual conditions, the water–gas shift reaction ($CO + H_2O \rightarrow CO_2 + H_2$) occurring in the MPR would provide additional hydrogen, thereby shifting the calculated hydrogen curve toward higher hydrogen/waste ratios. Thus, the operating point is the intersection of the carbon conversion curve and the appropriate hydrogen production curve. As shown in Fig. 5, the appropriate hydrogen production curve for a given gasifier depends on the amount of water–gas shift occurring as well as on the carbon in the char fed to the gasifier.

2. Effect on Product Gas Composition   A critical question and one of the key objectives of this study was to determine the yields, composition, and suitability of the raw product gas for SNG production.

As the tabulated results indicate, a considerable fraction of the carbon converted goes to methane with significant amounts of ethane and propane. For example, for both moving-bed and free-fall experiments the total gas phase hydrocarbons ranged from 40 to 50% of the total carbon in the gas phase. Carbon oxides are inevitable because of the relatively high oxygen content typical of solid wastes. Since these tests were made with a hydrogen feed gas rather than synthesis gas, the presence of carbon oxides in the product gas indicates substantial oxygen removal by carbon. At the relatively low pressure used, the concentrations of benzene are about what would be expected from its vapor pressure.

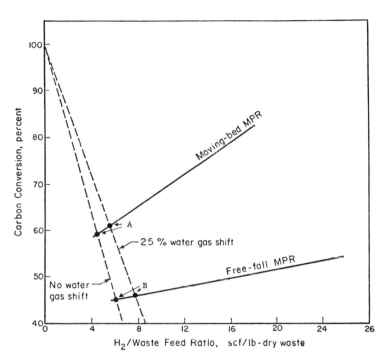

Fig. 5. Effect of contacting scheme on operating point in an integrated waste gasification system. ———, MPR operating lines; – – –, entrained gasifier operating line. A and B are operating points for integrated system.

Plots of the methane production per unit of waste fed and the hydrogen consumption per unit of methane formed are shown as functions of the hydrogen/waste feed ratio in Fig. 6 for both the moving-bed and free-fall reactor systems. These data indicate that for a given methane yield both a lower hydrogen/waste feed ratio and lower net hydrogen consumption are required in the moving-bed compared to the free-fall reactor system. Thus, higher hydrogen utilization efficiencies are achievable when operating in the moving-bed mode.

As the hydrogen/waste feed ratio decreases, the concentration of methane plus ethane in the product gas increases as indicated in Fig. 7, showing that at the operating condition of an integrated reactor system (cf. Fig. 1) the methane content of the product gas will be comparable to that from a Lurgi gasifier even after taking into account the effect of using synthesis gas rather than undiluted hydrogen to feed the MPR.

For example, using free-fall Run 6 data as a basis together with rather modest extrapolations of the experimental hydrogen/waste feed ratio used

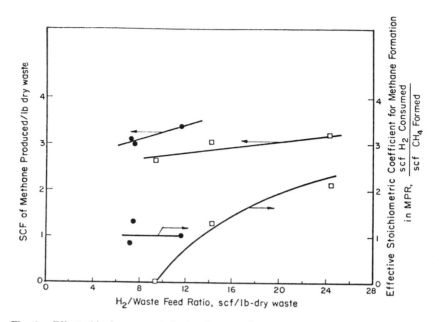

Fig. 6. Effect of hydrogen waste feed ratio on methane produced in MPR. Moving-bed (●) and free-fall (□) experiments.

in Run 6 to the hydrogen/waste ratio at the operating point (no water–gas shift) allows the small reduction in carbon conversion in the MPR to be predicted.

Assuming the same yield per pound of carbon and composition of synthesis gas reported by VonFredersdorff and Elliott [2], the composition and yield raw gas from the MPR operating in the free-fall mode and integrated with an entrained gasifier may be calculated. Similar calculations were also made based on moving-bed data (Run 10), and both the free-fall and moving-bed gas yield and composition data are summarized in Table III.

Thus, gas from either free-fall or moving-bed operation can be used to produce SNG after water–gas shift to adjust the H_2/CO ratio, acid gas removal, and methanation. The fraction of the total methane produced in the MPR approximately doubles in going from a free-fall to a moving-bed contacting system. It is not surprising that at the increased solid residence time in the moving bed hydrogen utilization efficiency is increased and more methane is produced directly than in the free-fall case. However, as has been previously pointed out, there are indications that in the free-fall case the solid waste never completely heats up to reaction temperature. Since in an integrated reactor system the hot synthesis gas from the

TABLE III

Calculated Gas Composition and Yield from Integrated Reactor System

	Free fall (run 6)	Moving bed (Run 10)
H_2	31.9	13.3
CH_4	10.4	17.2
CO	45.9	51.9
CO_2	10.1	16.1
C_2H_6	1.2	1.1
C_6H_6	0.5	0.4
Gas yield, scf/lb dry waste	19	16
Fraction of total methane produced in the MPR	0.26	0.52

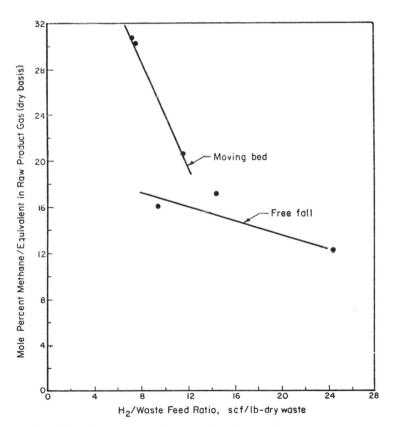

Fig. 7. Effect of hydrogen/waste feed ratio on hydrocarbon content of product gas.

entrained reactor will enter the MPR at about 2000°F there is a great probability that the free-fall operation in a commercial plant will give results closer to those from the moving bed.

In any case, the results achieved are comparable to those from coal gasification in the reactor systems intended to maximize direct methane formation such as Lurgi or Synthane.

V. CONCLUSIONS AND RECOMMENDATIONS

The experimental work reported herein establishes that a raw gas containing a substantial methane content can be continuously produced from solid waste under conditions that can be achieved in existing commercial equipment. Because of the short residence time required to effect the desired conversion levels the reactor systems can be quite compact. Scale-up should be no problem because the conversion rate seems limited by heat transfer which should occur at a higher rate in a commercial reactor because the heat source will be 2000°F synthesis gas.

Because free-fall operation will eliminate the possibility of plugging of solid waste, it should be the first system to be demonstrated. Later, the free-fall system can be converted to a moving-bed system by the installation of a moving grate.

Demonstration and commercialization of this process can, therefore, be quickly implemented to alleviate both solid waste disposal and natural gas shortage problems in urban areas.

ACKNOWLEDGMENT

Dr. Satya P. Chauhan provided the TGA data reported in Fig. 3.

REFERENCES

1. H. F. Feldmann, Pipeline gas from solid wastes, *AIChE Symp. Ser.* **68,**125–131 (1972).
2. C. F. VonFredersdorff and M. A. Elliott, Coal gasification, *in* "Chemisty of Coal Utilization" (H. H. Lowry, ed.), Supplementary Vol., Chap. 20, pp. 982–983. Wiley, New York, 1963.

Chapter V

THE NATURE OF PYROLYTIC OIL FROM MUNICIPAL SOLID WASTE

Kenneth W. Pober and H. Fred Bauer
OCCIDENTAL RESEARCH CORPORATION
LA VERNE, CALIFORNIA

I. INTRODUCTION

Several years ago, Occidental Research Corporation (formerly Garrett Research and Development Company) initiated a coal research program to explore the development of a process that would provide for the economic conversion of coal reserves to synthetic liquid fuels. The result was a Flash Pyrolysis™ process with the following main features: near ambient pressure, no requirement for chemicals or catalysts, low capital investment, flexibility of feedstock, variability of temperature, and minimum feed pretreatment. It became apparent shortly thereafter that a similar process could be used to produce synthetic fuel oils from organic solid wastes. Synthetic fuels have since been produced from such diverse feedstocks as municipal refuse, tree bark, cow manure, rice hulls, grass straw, sewage sludge, and used tires.

The production of liquid fuels from waste is desirable for several reasons. A low sulfur, low ash fuel can be made for commercial use. Because the fuel is liquid it can be stored and transported. Thus the energy product does not have to be used at or near the recycling plant site. Further, it can be used in plants which are not continuously operating. This flexibility, plus the ability to sell the fuel on a calorific content basis, provides powerful use incentives for the community and customer alike.

Also under development today are a number of solid waste pyrolysis processes which maximize gas yields rather than liquids. Both a low-BTU and a medium-BTU gas production system have been discussed in Chapters II and III.

II. THE PROCESS

Flash Pyrolysis technology has been used with modification to process municipal and industrial solid wastes. A process schematic which incorporates the following operations is shown in Fig. 1:

(1) primary shredding of the raw refuse to -3 in.,

(2) magnetic separation of ferrous metals,

(3) air classification to remove most of the inorganics from the pyrolysis feed material,

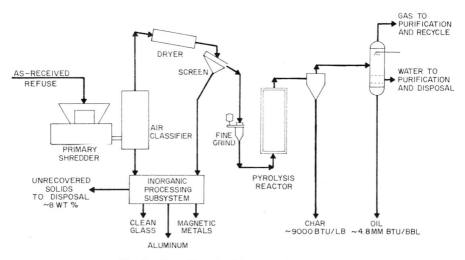

Fig. 1. Recycling of solid wastes via pyrolysis.

(4)   drying of the shredded refuse to about 3% moisture,

(5)   screening of the dry material to reduce the inorganics content to below 4% by weight,

(6)   recovery of aluminum, and a clean glass product,

(7)   secondary shredding of the organics to −14 mesh,

(8)   Flash Pyrolysis processing of the organics, and

(9)   collection of the pyrolytic products under rigid pollution control conditions.

The key function of feed preparation is to provide a dry, finely divided, essentially inorganic-free feed material suitable for the pyrolysis reactor. The effort spent in meeting this requirement also provides for the economic recovery of magnetic metals, aluminum, and clean glass. It is important to minimize the inorganic content in the pyrolysis feed material. Although the pyrolysis itself is virtually unaffected by the inorganics, these materials will degrade the quality of the final product oil, and increase maintenance costs for secondary shredding.

The Flash Pyrolysis process involves the rapid heating of finely shredded organic materials in the absence of air using a proprietary heat exchange system. The incoming feed is heated by contact with recycled char and pyrolyzed to form liquids, gas, and char. The char is separated by cyclones, the liquid is separated with a quench fluid, and the gas is recycled into the pyrolysis loop. Pyrolysis occurs at very short residence times (seconds) and high heat-up rates. This method maximizes the volatile yield and protects the products from further cracking. In the case of municipal solid waste, the emphasis is on liquid yield. A useful temperature range for liquid yield is 850 to 1050°F. A desirable fluidity in the product oil may be achieved by varying the moisture content. A moisture level of 14 to 18% gives a suitable viscosity for handling of municipal solid waste oil.

During laboratory pyrolysis studies using a small continuous 5-lb/hr reactor, oil yields of about 40 wt % were obtained with dried municipal wastes from which about 90% of the inorganics had been removed. These oil yields have been confirmed on a larger scale using a 4-ton/day pilot plant. Several other waste feedstocks were then examined in this process. Animal waste, rice hulls, fir bark, and grass straw have been pyrolyzed under conditions very similar to those employed for municipal waste. The data in the following tables are from typical experiments but are not necessarily optimum with regard to liquid yield and fuel value.

The analyses of the feedstocks are shown in Table I. The feedstock from municipal solid waste contains only 6.7% ash, 0.2% chlorine, 0.2% sulfur, and 0.7% nitrogen. It is a highly oxygenated material with a C/O ratio of 1.39. The other feedstocks are also highly oxygenated but they contain more

TABLE I

Analysis of Organic Solid Waste Feedstocks

Analysis (wt %)	Animal waste	Rice hulls	Fir bark	Grass straw	Municipal solid waste
C	39.3	39.4	48.3	45.0	44.2
H	4.7	5.5	5.3	6.0	5.7
N	2.3	0.5	0.2	0.5	0.7
S	0.6	0.2	0.0	0.5	0.2
O	28.1	36.1	34.3	42.0	42.3
Cl	1.7	0.2	0.2	0.4	0.2
Ash	23.3	18.2	11.7	5.7	6.7

undesirable components, particularly animal waste, which has 2.3% nitrogen, 1.7% chlorine, and 23.3% ash.

III. PRODUCT OIL CHARACTERISTICS

The pyrolytic oil is the most important product obtained from this recycling process. A comparison between the typical properties of No. 6 fuel oil and pyrolytic oil is shown in Table II. As might be expected from its genesis, pyrolytic oil differs in many important respects from fuel oil

TABLE II

Typical Properties of No. 6 Fuel Oil and Pyrolytic Oil

	No. 6	Pyrolytic oil
Carbon (wt %)	85.7	57.5
Hydrogen (wt %)	10.5	7.6
Sulfur (wt %)	0.7–3.5	0.1–0.3
Chlorine (wt %)	—	0.3
Ash (wt %)	<0.05	0.5–1.0
Nitrogen (wt %)	} 2.0	0.9
Oxygen (wt %)		33.4
BTU/lb	18,200	10,500
Specific gravity	0.98	1.30
lb/gal	8.18	10.85
BTU/gal	148,840	113,910
Pour point (°F)	65–85	90
Flash point (°F)	150	133
Viscosity SSU @ 190°F	340	3,150
Pumping temperature (°F)	115	160
Atomization temperature (°F)	220	240

derived from petroleum. As Table II shows, a gallon (or a barrel) of oil derived from the pyrolysis of municipal waste contains some 75% of the heat energy available from No. 6. Pyrolytic oil is also far more viscous than a typical residual oil. This means that the oil must be stored, pumped, and atomized at somewhat higher temperatures.

A. Combustion Characteristics

Combustion tests completed with some 230 gal of oil at the research facilities of a large manufacturer of power plant equipment have shown that (a) the pyrolytic oil could be pumped without trouble at 160°F, and (b) satisfactory atomization was achieved with 50 psi steam when 10 gal/hr of oil was delivered to the burner tip at 25 psi and 240°F. The pyrolytic oil has also been found to be compatible with several common sources of No. 6 fuel. Satisfactory combustion tests have been conducted on blends of 25 and 50% pyrolytic oil with No. 6 oil.

Ignition stability with the pyrolytic oil and with the blends was equal to that obtained with No. 6 alone; and stack emissions when burning pyrolytic oil and blends indicated negligible amounts of unburned carbon at excess oxygen levels over 2%.

The successful combustion of pyrolytic oil blended with No. 6 fuel oil was unexpected since the two are not miscible. The blend is a dispersion. Firing such a blend in a utility boiler has two important advantages. Our laboratory studies have shown that blends have a greatly diminished corrosive effect, beyond any dilution factor, on mild steel compared to pyrolytic oil alone. Second, pyrolytic oil can serve to "blend down" the sulfur content of a No. 6 oil which would otherwise be unacceptable.

The ash content is of particular importance to the end user. It affects the combustion operation. In contrast to solid, refuse-derived fuels, which have ash contents of at least 10% synthetic liquid fuels can be produced at under 1% ash, and can therefore be burned in power stations without ash-handling capability. An ash analysis of the oil shown in Table III is given in Table IV. Sodium and potassium are usually high, followed by iron and aluminum; the zinc value is abnormal. Up to 50% of the final ash recovered is water soluble.

B. Physical Properties

Pyrolytic oils derived from cellulosic feedstocks are complex fluids, as anticipated. For example, Fig. 2 shows a broad range of molecular weight for bark pyrolytic oil. Oil from municipal solid waste consists of a large number of compounds with an apparent molecular weight range of 32 to 10,000 and a boiling range of 55 to 300°C. Only half of this material is capable of being distilled. Oxygen content is always present so that the

TABLE III

Chemical Analysis of a Typical Solid Waste and Pyrolytic Oil

Component	Solid waste (wt %)	Pyrolytic oil (wt %)
Carbon	46.0	56.8
Hydrogen	6.4	7.6
Sulfur	0.12	0.2
Nitrogen	0.9	1.1
Chlorine	0.23	0.02
Moisture	2.6	6.4
Ash	5.6	0.32
Oxygen (by difference)	38.15	27.6

carbon-to-oxygen ratio (C/O) ranges from 1 to 2 for most fractions. The oil has a pungent, heavy barbecue odor, and a dark red-brown color.

Pyrolytic oils from cellulosic feedstocks have densities 30 to 50% greater than fuel oils. The densities of municipal waste, fir bark, and rice hull oil versus temperature are shown in Fig. 3. It can be seen that rice hull oil is inherently less dense than either fir bark or municipal waste oil. The dependence of density on moisture content was determined at room temperature for fir bark and municipal waste oil, and is shown in Fig. 4.

TABLE IV

Ash Analysis of Pyrolytic Oil from Municipal Solid Waste

Element	Oil (wt %)	Ash (wt %)
Ash	0.32	—
Zn	0.086	28.7
Cu	0.004	1.3
Al	0.005	1.7
SiO_2	0.006	2.0
Mg	0.004	1.3
Fe	0.010	3.3
Ca	0.010	3.3
Na	0.065	21.7
K	0.034	11.3
V	0.001	0.3
Ti	0.001	0.3
Total	0.226	75.2

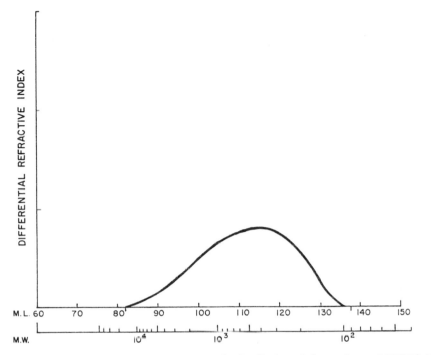

Fig. 2. Bark pyrolytic oil molecular weight distribution. Column, Styragel X17712-49; solvent, THF; number average molecular weight, 365; weight average molecular weight, 786.

The viscosity of municipal solid waste oil is shown in Fig. 5. Typical values are somewhat higher than most No. 6 fuel oils. A family of curves at five different moisture content values is plotted as a function of reciprocal absolute temperature. Viscosity is also a function of thermal history, since reheating the oil for extended periods above 180°F causes some degradation. The resultant thickening is less severe in air than under a N_2 or CO_2 blanket. In general, the fluids are non-Newtonian, showing a hysteresis especially at lower temperatures and a shear-dependent viscosity.

The acidity of pyrolytic fluids is due primarily to carboxylic acids. The oil has a pH of 2.2 to 3.0 measured in either water or methanol, and owes its acidity to the presence of formic, acetic, glycolic, malic, and acrylic acids. Corrosion of pretreated 304 stainless steel at oil pumping temperatures, 130–140°F, has been measured at 0.05 to 0.10 mil/year for municipal solid waste oil. As expected, carbon steel corrodes an order of magnitude faster and 316 stainless steel slower by a factor of 10. Pyrolytic oil is similar to acetic acid in terms of corrosive behavior. Corrosion rates are substantially reduced by blending municipal solid waste oil with No. 6 fuel oil.

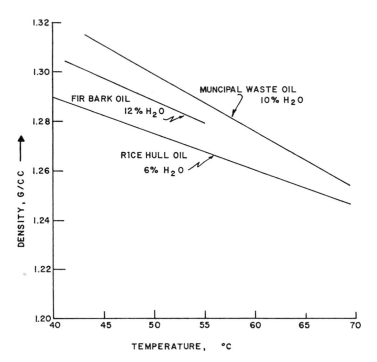

Fig. 3. Fluid density versus temperature.

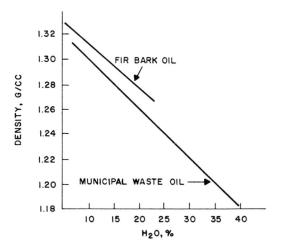

Fig. 4. Fluid density versus moisture. Temperature 27°C (80°F).

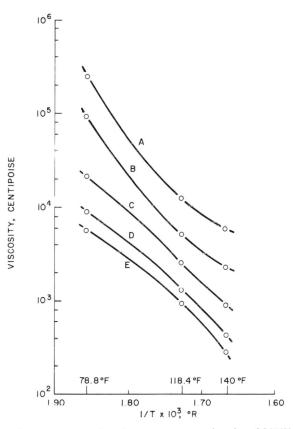

Fig. 5. Effect of temperature and moisture content on viscosity of MSW pyrolytic oil—Run 61-73, Drum 473. % H_2O for curve A, 8.56; B, 11.15; C, 13.30; D, 14.97; E, 16.79.

Solvents for pyrolytic oil are dilute, aqueous caustic, oxygenated solvents such as diethylene glycol, cellosolves and carbitols, and very polar solvents such as dimethyl formamide and dimethylsulfoxide. Benzene, hexane, and ether are very poor solvents (0–10% solution). Water and pyrolytic oil from municipal solid waste are miscible up to a 4:1 ratio of water to oil. Beyond this point, solids will precipitate, whereas 60% of the oil remains water soluble.

C. Chemical Composition

Pyrolytic oils derived from solid waste may be fundamentally similar to pyrolyzed cellulose. Pure cellulose, chains of glucopyranose, depolymerizes and rearranges at elevated temperature to form levoglucosan, which undergoes subsequent fragmentation and polymerization. The extent to which municipal solid waste follows this scheme is speculative, since a number of factors are involved. Noncellulosic materials (degraded plastics, ash, metals, etc.) in solid waste may either catalyze or suppress competing reactions. Our reaction time and temperature are different from conditions reported elsewhere for cellulose pyrolysis work. Variation in plastic content, vegetable waxes, lignins, and ash composition causes some variation in yield, fuel value, and viscosity.

A chemical analysis of a typical pyrolytic oil is given in Table II. The composition of the solid waste feed is also shown. The empirical formula best fitting the oil analysis is $C_5H_8O_2$. Since the oil was derived primarily from cellulose, $C_6H_{10}O_5$, it is reasonable that the first two steps occurring during Flash Pyrolysis processing are the loss of CO_2 and H_2O. Further splitting occurs at longer residence times, and yields a more viscous tar but greatly reduces oil yield.

Solubility classes are described in Fig. 6. Employing a classical solubility separation scheme, pyrolytic oil from municipal solid waste consists of two main fractions: a water- and ether-insoluble, aqueous base-soluble fraction (pH 8), and a water-soluble but ether-insoluble fraction. These two fractions comprise 35 and 58 wt %, respectively, of pyrolytic oil on a dry basis. Both fractions were quantitatively analyzed for various functionalities such as carboxylic, carbonyl, alcohol, olefin, and ether groups. Functional groups and their relative proportions in pyrolytic oil from municipal refuse are shown in Table V. The complex nature of the oil is apparent, and is confirmed with high pressure liquid chromatography and gas chromatographic analyses.

TABLE V

Functional Groups in Pyrolytic Oil from Municipal Waste

Functionality	Carboxylic acid fraction		H_2O-soluble–ether-insoluble fraction	
	Equivalent weight	Relative amount	Equivalent weight	Relative amount
Hydroxyl	250	1.5	160	3.1
Carbonyl	230	1.6	140	3.6
Carboxyl (neutral eq.)	375	1.0	500	1.0
Ether	120[a]	3.1	>350[c]	1.4
Olefin	b		340	1.5

[a] Calculated by difference, assuming fraction 35.3 wt % oxygen.

[b] Unable to find bromination solvent; H_2O, CCl_4, DMSO, pyridine cannot be used.

[c] Calculated assuming dry pyrolytic oil was 30 wt % oxygen.

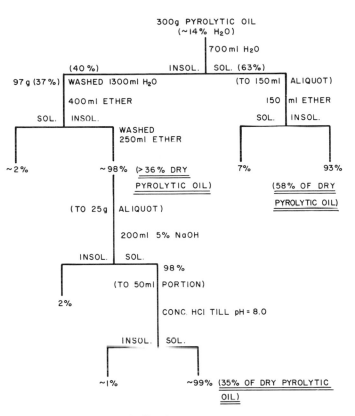

Fig. 6. Pyrolytic oil solubility classes.

Characterization is made more manageable through solvent refining. In a large excess of water, pyrolytic oil will partition into a 40% yield of solids and 60% of water-soluble oil. The recovered, light brown solid, like the oil, is very complex. The yield, melting point, and thermogravimetric data for the solid vary with solvent composition, and rate and temperature of mixing used to obtain the solid. Molecular size separation achieved with a set of gel permeation columns showed that the solid consisted of a group of four to five oligomers in the 100 to 1000 molecular weight range. Carbonyl and hydroxy groups are present, ring opening has probably occurred, but the molecular weight is less than 1000. Thus, Flash Pyrolysis processing of municipal solid waste does not result in other cellulosic or modified high polymers. Much lower molecular weight, rearranged materials are produced. Identification of these materials is underway. Pyrolysis oil itself shows a similar analysis pattern, though it is slightly more complex as expected.

The water-soluble fraction of pyrolytic oil can be concentrated or solvent extracted. Over 25 compounds are present. Formic and acetic acids, methanol, and acetone have been identified. The molecular weight of the components in this fraction is less than 150, and the yield of any individual component is low. The origin of these products is difficult to identify, since they can be noncellulosic or represent secondary and tertiary pyrolysis products.

IV. CHAR AND GAS PRODUCTS

All pyrolysis processes produce gaseous and solid products as well as liquids. These by-products also contain useful energy—either for use in plant operations or for export and sale. The by-product yield, dependent on both the feed material and the operating conditions, is highly important to the overall operation of the waste-to-energy system.

The char yields and analyses are shown in Table VI. Yields of char vary from 20% for municipal solid waste to 48% for animal waste. Heating values range up to 9000 BTU/lb for char from municipal solid waste, similar to the heating value for subbituminous coal. All the chars contain large amounts (21–49%) of ash.

The gas analyses from pyrolysis of solid wastes to liquids are shown in Table VII. They are low heating value gases, from 200 to 500 BTU/scf, because they contain from 27 to 65% CO_2. They contain from 1.7 to 7.8% H_2S except for municipal solid waste gas which contains 0.4% H_2S.

The distribution of the products—oil, char, gas, and water—has a significant effect on the economics of the overall process. The gas produced under

TABLE VI

Pyrolytic Char from Organic Solid Wastes

Analysis (wt %)	Animal waste	Rice hulls	Fir bark	Grass straw	Municipal solid waste
C	34.5	36.0	49.9	51.0	48.8
H	2.2	2.6	4.0	3.7	3.3
N	1.9	0.4	0.1	0.5	1.1
S	0.9	0.1	0.1	0.8	0.4
O	7.9	11.5	24.3	19.2	12.8
Cl	3.7	0.2	0.2	0.5	0.3
Ash	48.8	49.2	21.4	24.3	33.3
BTU/lb	5450	6100	8260	8300	9000
Char Yield (%)	48.1	35.9	41.6	23.2	20.0

TABLE VII

Pyrolytic Gas from Organic Solid Wastes

Analysis (wt %)	Animal waste	Rice hulls	Fir bark	Grass straw	Municipal solid waste
H_2	6.1	23.3	6.3	9.6	11.7
CO	21.9	28.5	14.7	53.9	34.9
CO_2	55.9	40.8	64.7	26.9	35.4
CH_4	6.1	3.5	10.5	3.8	5.7
C_2H_4	0.5	0.8	1.0	0.8	4.0
C_2^+	1.8	0.8	1.2	2.8	2.4
H_2S	7.8	2.4	1.7	2.2	0.4
BTU/scf	226	214	222	349	350–500
Gas Yield (%)	10.8	12.1	8.2	5.2	30.0

optimum liquefaction conditions has a moderate heating value of about 500 BTU/ft at the outlet of the reactor. This fuel is burned on-site for process heat and a portion of the char is used for the same purpose.

V. CONCLUSION

The Flash Pyrolysis process affords an effective means of converting municipal solid waste to fuel. Over one barrel of pyrolytic oil can be obtained from each ton of as-received refuse. Since the original research, an integrated solid waste recycling process has been developed which can convert or reuse about 90% of the raw materials obtained in municipal refuse.

A vital aspect of any resource recovery system is the state of purity of the products to be returned to the economy. Initial resistance to recycled materials is largely due to the quality of the recovered products, rather than a lack of suitable markets. A substantial effort has been spent on upgrading the quality of the recovered materials—glass, metals, and oil.

It remains to demonstrate this Flash Pyrolysis process and the concept of resource recovery for the public, and this is being done in San Diego. A 200-ton/day demonstration plant is under construction with start-up scheduled for mid-1977. The plant is a joint effort of Occidental Petroleum Corporation, the U.S. Environmental Protection Agency, and San Diego County. San Diego Gas and Electric Company will burn 200 barrels/day of pyrolytic oil for power generation. Flash Pyrolysis technology offers a flexible and economical means of converting large quantities of solid waste into a solution for the nation's expanding energy requirements.

Chapter VI

THE CONVERSION OF FEEDLOT WASTES INTO PIPELINE GAS

Frederick T. Varani and John J. Burford, Jr.*

RESEARCH DIVISION
BIO-GAS OF COLORADO, INC.
ARVADA, COLORADO

I. INTRODUCTION

Bio-Gas of Colorado has been dealing with the treatment and reuse of agricultural waste organic materials for two years. Research and studies have been conducted in Colorado, Arizona, Utah, and New Mexico in conjunction with grants from the 4-Corners Regional Commission. Most of the

* Present address: Bio-Gas of Colorado, Inc., Loveland, Colorado.

design information presented here has been accumulated from pilot plant operations and studies at the Monfort of Colorado, Inc., beef feedlots.

A. Agricultural Waste Definition

Agricultural waste material such as steer manure has an energy value which ranges from 3000 to 8000 BTU/lb of solid material. This material is currently being used, almost exclusively, in its traditional manner as an addition to agricultural croplands. Various processes, well known and reported, are available to convert this manure into a form of energy product such as oil or gas. The most often reported processes are (1) direct combustion, (2) pyrolytic conversion, and (3) anaerobic digestion.

Agricultural wastes of an energy-producing nature are defined as (1) residue from animal husbandry operations, (2) food processing wastes, and (3) crop residue left in the fields (or collected). These materials are all generated in large quantities. Items 1 and 2 are in a reasonably centralized or collected form and are generated on a continuous basis. Item 3 is neither generated continuously nor normally collected.

In the Southwest, most crop residues are not removed from the fields and most agriculture state agents strongly recommend leaving this material on the fields for erosion control in the windy climate.

B. Amounts of Animal Waste Generated

Anderson [1] estimates that 210 million tons of moisture- and ash-free organic material from animal husbandry operations are generated yearly in the United States.

Burford and Varani [2] estimate that 938,398 tons/year of dry organic material (volatile solids) are collectible from 17 areas of concentrated livestock feeding in Colorado. The maximum distance considered for transporting manure was 15 miles. This information was generated during the summer of 1975 and represents approximately 52% utilization of the state feedlot and animal husbandry capacity.

In Colorado, eight areas or potential sites have been located in which the quantity of manure generated is large enough to justify a utility size methane facility. These areas are located on maps published by the 4-Corners Regional Commission in September 1976, and the estimated quantities for Colorado are shown in Table I.

C. Potential Impact

In the final report to the 4-Corners Regional Commission, the authors estimated the amounts of manure and other agricultural wastes which are

TABLE I

Colorado Manure Inventory and Gas Prediction

	Cattle	Dairy	Hogs	Sheep	Chickens	Turkeys	Paunch manure
Livestock basis (Head or kill/day)							
Livestock in all site areas	767,248	36,050	104,000	201,240	1,457,810	3,178,600	4900
Livestock in potential utility site areas	641,472	22,100	80,000	98,120	1,420,580	3,141,800	4030
Solids per animal (lb of VS)							
Expected volatile solids scraped per animal per day	5.33	7.02	0.61	0.66	0.06	0.14	14.40
Methane potential (MCF CH_4/day)							
In all site areas	19,425	380	381	199	350	1780	353
Gas potential in utility site areas	16,241	233	293	196	341	1759	290
Total gas potential							
In all site areas	22,868						
In all utility site areas			19,353				
In utility site areas with collectable factor applied[a]					17,672		

[a] Cattle, 100% manure from lots of 1000 head or greater, and packing house waste; 60% from cattle feedlots less than 1000 and dairy waste; 50% from hog, sheep, chicken, turkey, and cotton gin operations are collectible.

located in the states of Colorado, Utah, New Mexico, and Arizona and have arrived at the following conclusion [2]:

The quantities of gas predicted in the utility site areas with a collectable factor applied represent 7.3% of the 1974 residential consumption for Colorado, 12.6% for Arizona, and 7% for New Mexico. These same gas quantities represent 2.2% of the total 1974 market consumption for Colorado, 2% for Arizona, and 0.7% for New Mexico. These gas figures are a result of the current, June, 1976, levels of cattle feeding for Colorado and Arizona. The New Mexico inventory is current as of January, 1976 and Utah as of January, 1975.

To say that this was the ultimate potential gas production available from *bio-conversion* would be misleading. This study has not considered the potential available from municipal refuse which provides a viable candidate material. The agriculture wastes mentioned thus far do offer sufficient quantities of supplemental natural gas to justify specific plant implementation and thus a *starting point* for a bio-conversion industry. To put the production quantities mentioned into perspective, the Colorado production of 17.67 million cubic feet of methane per day could support the connection of 44,175 new homes in Colorado at an average consumption of 400 cubic feet of gas per day. This approximately represents the demand for new connections in Colorado for one year. Admittedly a drop in the bucket of overall state consumption, but nonetheless significant to those who would receive the connections.

II. PROCESS SELECTION

Manure, as delivered, from an open dirt feedlot (the most common type of operation found in Colorado or the Southwest in general) would be extremely variable in (a) moisture content, (b) dirt or grit ratio, and (c) extent of decomposition caused by exposure during the confinement period. Figure 1 shows decomposition and organic carbon loss in a typical manure sample versus exposure time. Figure 2 shows a typical steer manure as received per ton breakdown.

A. Efficiency Considerations

Due to the nature of the feedlot manure, the process of anaerobic digestion has been the only process studied for feedlot manure conversion. Manure as a "fuel" must first be dried and further processed before direct combustion or pyrolytic decomposition could be practical. However,

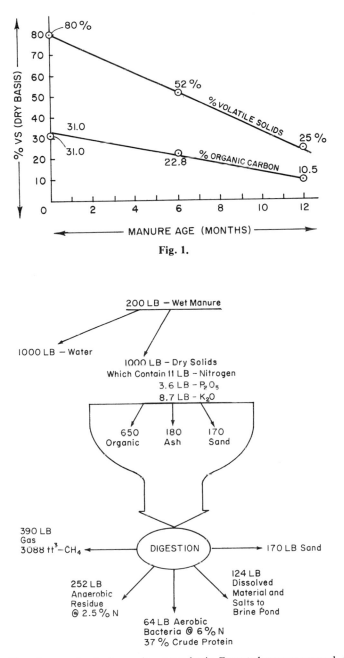

Fig. 1.

Fig. 2. Plant input and output on unit wet ton basis. Expected manure scraped, 4.0 tons per head annually.

anaerobic digestion is not adversely affected by moisture content, being a process whereby dilution with water is accomplished before utilization. Dilution allows the grit to be removed by simple sedimentation. The other mentioned processes all require a drying operation before combustion or pyrolysis can take place and this drying would place these processes at an energy disadvantage with respect to anaerobic digestion.

Pyrolysis takes place at elevated temperatures and pressures, combustion at elevated temperatures. Anaerobic digestion takes place at near ambient temperatures and pressures. These two parameters place the equipment cost per unit of energy produced lower for the digestion system.

A disadvantage of the digestion process is the long residence time required for an economic yield. This requires large reactor vessels. Most research to date has centered on reducing the digestion time and/or the cost of the reactor vessels.

B.   Total Material Utilization

The most important reason that digestion is viewed by the authors as the *only* viable process for utilizing the energy value of the manure is that the process allows the nutrient and humus values of the manure to be available to the farm community. Any process that destroyed the nutrient and humus value of manure would be found in much disfavor by the agriculture community.

The sludge (digested residue of the process) contains most of the nutrient values originally available in the manure. This material is in a liquid form, can be easily applied to croplands, and being a processed material is more consistent in its values over a long time than is the original manure. Combustion or pyrolysis would destroy these values for the most part.

III.   THE ANAEROBIC DIGESTION PROCESS

Anaerobic digestion is a process which utilizes bacteria to decompose (ferment) the organic fraction of the waste material. Anaerobic fermentation is one of the major biological waste treatment processes employed for municipal wastewater treatment. A comparison of aerobic and anaerobic methods for treating sewage indicates that anaerobic digestion becomes more economical when the feed concentration of volatile solids is greater than 0.4% of the waste stream on a wet basis. Since sewage is a relatively dilute waste stream, a sewage treatment plant is usually a combination of aerobic treatment of the dilute materials and anaerobic treatment of the thicker slurries. Figure 3 illustrates a typical sewage treatment plant flow-chart.

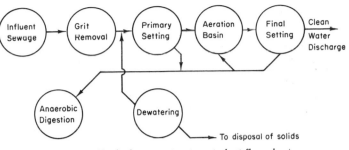

Fig. 3. Typical sewage treatment plant flow chart.

A. The Basic Process Description

From a biological viewpoint, anaerobic digestion may be described as a three-step process involving (1) hydrolysis of complex material, (2) acid production, and (3) methane fermentation. In the first step, complex organics are converted to less complex soluble organic compounds by enzymatic hydrolysis. In the second step, these hydrolysis products are fermented to simple organic compounds, predominantly volatile fatty acids, by a group of bacteria collectively called "acid formers." The dominant acids produced are acetic, propionic, and butyric, acetic and propionic being responsible for above 85% of methane produced. In the third step, the simple organic acids are fermented to methane and carbon dioxide by a group of substrate-specific, strictly anaerobic bacteria called the "methane formers." Thus, organic waste materials are converted to bacterial protoplasm, dissolved by-products, and gaseous end products (methane and carbon dioxide).

In this complex multistep process, the "rate limiting" step is usually this last step, the methane fermentation. This results from the 4- to 10-day regeneration times for the various species of methane-forming bacteria. Thus control of the anaerobic digestion process resolves to providing the optimum chemical and physical environment for the methane-forming bacteria, since these bacteria are the slowest to regenerate and are very sensitive to changes in their environment. Table II shows the significant parameters which must be controlled for successful anaerobic digestion.

Some of these factors are discussed briefly to demonstrate that, like any conventional chemical process, a selected operating point seeks to optimize certain performance parameters.

(1) *Oxygen-free environment* Those bacteria known as methanogenic bacteria are strict anaerobes and cease functioning in the presence of oxygen. This requires sealed tankage.

TABLE II

Physical and Chemical Factors

Physical factors	Chemical factors
Solids retention time (SRT)	pH
Oxygen free environment (absence of oxygen)	Alkalinity
Temperature	Volatile acid content
Solids concentration	Nutrients
Degree of mixing	Toxic materials
Solids loading rate	

(2) *Proper digestion time* The bacteria function at a rate proportional to temperature between 60 and 110°F (15.56–43.33°C) mesophilic range, and 120 to 150°F (48.89–65.56°C) thermophilic range. At any given temperature enough time (minimum digester tank volume) must be provided to allow the methanogenic bacteria to process the organic material properly. A minimum time of 10 days is required at the mesophilic temperature of 98°F and 4.50 ft^3 of methane is generated for every pound of organic matter introduced into the system.

(3) *Temperature uniformity* Although digestion will proceed at any of the temperatures mentioned, temperature changes greater than ±2°F in any 24-hr period are enough to cause "temperature shock," a phenomenon whereby the bacteria become relatively dormant and gas production ceases. This requires a temperature control system and insulation of the digestion vessel.

(4) *Nutritionally balanced feedstock* The bacteria require basically, organic carbon (lignin or nonorganic forms of carbon will not digest), nitrogen, phosphorus, and trace elements. Manure has enough of all the nutrients required. Increases in organic carbon alone could be tolerated with a resulting increase in gas production.

(5) *Absence of toxic elements* Heavy metals and ionic material of high concentration can cause bacteria to cease functioning.

When the proper conditions are provided, the bacterial action can take place, followed by the process of degradation or fermentation.

The gas released from the process is known as biogas and consists of roughly 50–70% CH_4, 30% CO_2, and a trace of H_2S by volume.

The pilot plant runs made by Bio-Gas of Colorado, to date, consistently produced a gas of 62–65% CH_4 by volume (at Denver altitude). They verified that 4.0–5.0 ft^3 of CH$_4$ could be generated for every pound of organic matter.*

The bacterial action is complex and a discussion of bacterial fermentation is the intent of this chapter. However, for those interested, a bibliography of excellent sources for this discussion is included in the reference section of this chapter [3–6]. A more detailed explanation of anaerobic digestion process kinetics may be found in the literature [7–11].

B. Research and Demonstration Programs

To date many researchers from various private and public organizations have recognized the potential of a bioconversion industry.

The Institute of Gas Technology has been conducting research on the bioconversion process for many years, both on agricultural and municipal wastes. In the spring of 1976, the U.S. Energy Research and Development Administration awarded a research contract to Waste Management Inc. of Oak Brook, Illinois to build a 100-ton/day municipal trash anaerobic digestion facility. This facility will be built at Waste Management's Pompano Beach, Florida facility and will be designed to answer the scale-up considerations arising from the research done to date on municipal refuse. Many agricultural colleges and the U.S. Department of Agriculture have active projects involving the digestion of wastes from animal husbandry operations. One such pilot plant is being located at the Agricultural Research Center in Clay County, Nebraska. It will process cattle wastes initially, then it will focus on hog wastes and the blending of animal and crop wastes.

In other research, efforts have been made to capitalize on the naturally occurring anaerobic activity present in landfill operations. Sanitary landfills in San Francisco and Palos Verdes, California have been drilled for the purpose of extracting usable volumes of methane.

Outside the United States small-scale digesters have been developed. Much of this effort began with the Gobar Gas Plant fabricated at the New Delhi, India, Agricultural Research Institute in 1939. China now has half a million small-scale digesters, India is installing some 100,000 such plants, and Korea is building 50,000 small-scale anaerobic operations. Some have been built on U.S. farms as well. However, these do not produce commercial quantities of methane. They generate only enough for the plant operator.

---

* On a theoretical basis the breakdown of 1 lb of cellulose would yield 7 ft^3 of methane; 4.50 ft^3 is an actual yield.

C. Full-Scale Commercial Plant Considerations

Bio-Gas of Colorado began in 1973 to characterize the waste from cattle feedlots and dairies and to develop a plant flow schematic which would allow "commercial" quantities of gases to be evolved.

Figure 4 shows a basic flow schematic as proposed for the Monfort Gilcrest feedlot. This facility is sized to handle 100,000 cattle units of manure input. (A cattle unit is one animal of 1000 lb weight.) Again a complete description of the process (one of many proposed) is beyond the scope of this chapter, but the basic steps of the process are as follows:

(1) The feed-manure is mixed with water.
(2) Sand and grit are removed.
(3) The resulting slurry is heated.
(4) The heated slurry is introduced into digestion vessels where digestion occurs.
(5) Biogas is removed, and H_2S and CO_2 are removed, compressed, and the purified CH_4 is sent into the interstate pipeline.
(6) Residue is removed from the digestion vessels, solids are separated, and the remaining liquid is admitted to the aeration basin.
(7) The liquid is aerated, allowing aerobic bacteria to grow.
(8) Solids are again separated and the remaining liquid is remixed with manure.

1. Plant Performance A 100,000-cattle unit facility would input 800,000–1,200,000 lb of dry solids per day to the plant. The plant will use four digester vessels of 1,000,000 ft³ capacity each and will generate 2,000,000–3,400,000 ft³/day of CH_4 at a cost of $2.00 to $2.60/1000 ft³. The biogas would be cleaned to the extent necessary for pipeline use and compressed to a pipeline pressure of 850 psig for sale.

2. Capital Costs The engineering of this facility has been on-going for the past two years and latest capital cost estimate shows a $3,220,000 to $6,500,000 cost ($1610–$1910 capital cost per daily generated MCF). Figures 5 and 6 show capital cost figures for digestion systems of various sizes as explained on each figure. The system cost is related mainly to the gas production/solids input and is sensitive only to the throughput of the facility and the biological activity of the manure or waste handled.

Several of the smaller sizes as shown on Fig. 5 and in the photograph of Fig. 8 have been built. The statistics given in Fig. 5 include 40% markup over cost for the builders.

Figure 6 is estimated and assumes costs as related to an owner–operator; however, IDC and a contingency of 15% are allowed in Fig. 6. As can be seen, from 10,000 to 100,000 cattle units is the least size sensitive area on the

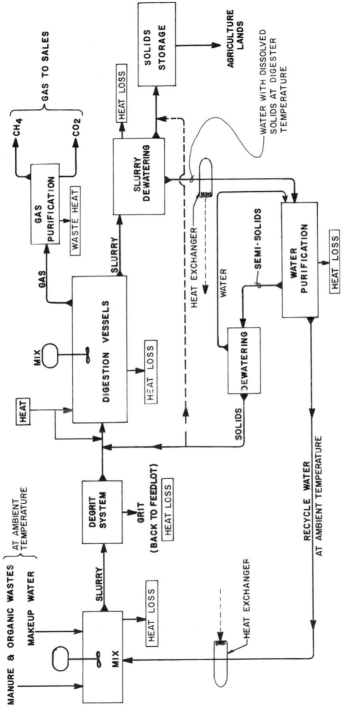

Fig. 4. Major block diagram agricultural waste/gas production.

97

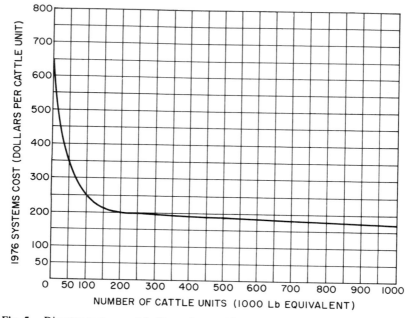

Fig. 5. Digester system cost (selling price to ultimate user). System based on conventional steel tank construction, including automatic manure processing and feeding, gas cleaning, compression and some storage, and liquid sludge disposal.

curve and 40,000 to 50,000 cattle units would be required to "make the deal" interesting to anyone contemplating a manure/gas facility.

3. Gas Production Cost　A sensitivity study has been generated focusing on the cost to produce gas with the facility described here. Table III shows the effect on gas cost per MCF by the respective cost factors for the 100,000 cattle unit plant.

The schedule in Table IV gives a breakdown of the gas cost in terms of major cost factors. The breakdown uses the 100,000 head cattle plant size. At this plant size, each $100,000 of annual operating cost affects the ultimate gas cost by 11.1¢.

As can be seen the largest increment on gas cost is the net cost for the money, followed by the cost for the fuel (i.e., manure). All other costs add an amount about equal to the two largest increments.

The "net" manure cost is assumed to mean cost invested in obtaining manure delivered to the plant minus revenues obtained by selling the sludge residue.

4. Construction Techniques　The process flow schematic is relatively simple. The facility consists basically of large tanks and lagoons for holding

TABLE III

Gas Production Unit Costs

Net manure cost	$0.28/MCF for every $1.00/ton net that must be paid for manure
Capital cost	$0.09/MCF for every $500,000 change in capital investment required
Increased manure input	$0.14/MCF less for every additional 50 tons/day available to plant for purchase
Volatile solid to methane conversion	$0.11/MCF added or subtracted to the gas cost for each 0.25 ft³/lb VS change in the gas yield from volatile solids

slurries and allowing anaerobic and aerobic bacteria to process the material feedstock. The physical size of the tanks required to allow the proper hydraulic detention time is the feature which causes the largest increment of capital cost in the envisioned facility. Traditional sewage plant design relies on concrete and/or steel tanks, each custom engineered and field erected.

A factor cost often reported for these municipal installations is $2.00/ft³ of digester volume.

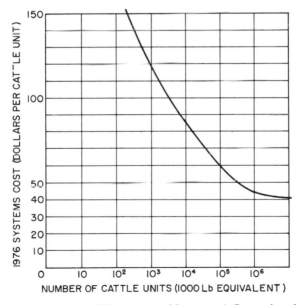

Fig. 6. Digester system cost (building cost to ultimate user). System based on rubber-lined trench construction, including automatic manure processing and feeding, gas cleaning, compressing and some storage, sludge dewatering and storage, and liquid recycle.

TABLE IV

Gas Cost Breakdown for a 100,000 Head Cattle Plant

Factor	Cost (¢)
Manure	55.5
Digester heating	8.4
Gas cleaning and compression	20.3
Labor	18.6
Other operating costs	12.1
Interest and depreciation	72.9
Other fixed expenses	21.4
Gross profit	42.4[a]
Total	251.6[b]

[a] Eight percent on capitalized investment.
[b] Per generated 1000 ft³.

In the envisioned facility, all tanks including the digester vessels themselves, the clarifiers, and lagoons are all "Hypalon-lined" in-the-ground trenches, a type of construction finding favor in more recent waste treatment projects. Use of this type of construction has allowed keeping the total capital costs under $2.00/ft³ for this type of facility. This includes the extra equipment such as slurry mixing, gas cleaning and compression, and liquid aeration that a municipal facility would not require.

Figure 7 shows plan and elevation views of the proposed digester construction.

5. *Digester Heating* Judicious use of insulation and heat exchange is required in the process to keep the net energy requirements as low as possible. A coal-fired boiler will provide a majority of the net digester heating requirement. Augmenting heat sources will be waste heat of compression and solar energy. A form of flat plate collector using digester effluent as the heating medium has been patented and incorporated into the system design.

D.   Mobile Pilot Plant

Figure 8 shows a mobile pilot plant built by Bio-Gas of Colorado for the 4-Corners Regional Commission with the various major features shown. This unit would be represented by the curve in Fig. 5. The unit shown is capable of handling the wastes from 10 to 30 dairy cattle, or 20 to 50 feedlot animals and can produce up to 1000 ft³ of methane a day. The various engineering statistics for the machine shown are presented in Table

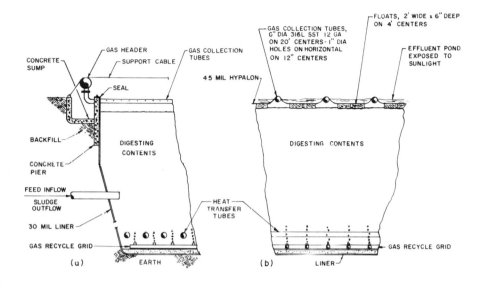

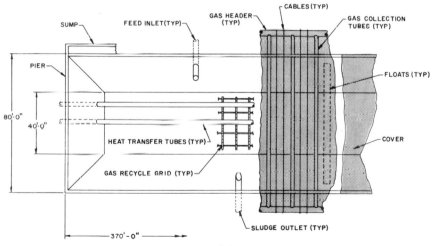

Fig. 7. Bio-Gas digester details. Cross section of (a) width and (b) length. (c) Partial top view.

TABLE V

Operational Statistics for Mobile Digestion Unit

Main tank	
Capacity	6000 gal
Feed rate	500 gal/day
Size	7 ft o.d. × 21 ft long, made from ¼ in. thick carbon steel
Insulation	4 in. sprayed urethane "1" factor at 7.1 per in.
Heater	40 ft², panel coil, 20 ft² per tank side
Circulation pump	⅓ hp centrifugal
Mix tank	
Capacity	500 gal
Feed rate	5 gal/min
Mixer	¾ hp Cleveland Mixer Corporation, rotates @ 420 rpm
Pump	Moyno "Mazerator" 1½ hp, positive displacement, high viscosity duty
Heat exchanger	20 ft² panel coil
Solar collector	
Size	140 ft²
Output	140,000 BTU (average Colorado day)
Weight	600 lb
Manufacturer	Rocky Mountain Heating and Sheet Metal, Denver, Colorado
Construction	Copper tubing soldered to copper sheet, backed with 2 in. urethane foam
Circulation pump	⅓ hp, centrifugal
Gas cleaning system	
Tower	Rated at 300 psi, plastic (fiberglass construction) having plastic packing. 8 in. o.d. × 6 ft overall height
Gas compressor	Manufactured by Corken, Inc., supplied by Air-Mac, Inc., of Denver. 2 hp, rated @ 3 cfm @ 100 psig
Circulation pump	½ hp, gear pump
Miscellaneous systems	
Sludge pump	½ hp, capable of solids handling
Generator	2.5 kVA, 5 hp gasoline (or methane) fueled engine
Temperature control systems	Jordan Valve (self-vapor-powered) temperature sensing and flow regulation valve
Electrical systems	All weather-tight construction

Fig. 8.

V. This machine is used as a demonstration for agribusiness application of the process described and also to verify "yield" numbers quoted herein.

IV.   CONCLUSION

A potentially viable energy-producing industry could be operated in Colorado and in other agriculturally oriented states for the following main reasons:

(1)   Manure is available in "commercial" quantities in a collected form.
(2)   Today manure is available at \$2.00/ton or less (\$0.615–\$0.705 per million BTU). As a fuel manure today is cheaper than coal.
(3)   The technology is well known to convert the energy value of manure into a very desirable form.
(4)   It appears that the cost of capital for the conversion facility (under \$2000 per generated MCF) is also desirable.

REFERENCES

1. L. L. Anderson, A Wealth of Waste; A Shortage of Energy," Chap. 1, this volume.
2. J. Burford and F. Varani, Energy potential through bio-conversion of agriculture wastes. 4-Corners Regional Commission, Final Rep. to 4-Corners Regional Commission, FCRC Grant No. 651-366-075, 1976.
3. W. J. Jewell, "Energy, Agriculture and Waste Management." Ann Arbor Science Publishers, Inc., Ann Arbor, Michigan, 1975.
4. P. L. McCarty, Anaerobic waste treatment fundamentals: I. chemistry and microbiology, pp. 9, 107. Public Works 95, 1964.
5. Metcalf and Eddy, "Wastewater Engineering." McGraw-Hill, New York, 1972.
6. F. G. Pohland and S. Shosh, Kinetics of substrate assimilation and product formation in anaerobic digestion, *J. Water Pollut. Control Fed.* **46** (4), 1974.
7. F. G. Pohland, "Anaerobic Biological Treatment Processes," Advances in Chemistry Series 105. American Chemical Society.
8. C. N. Sawyer and P. L. McCarty, "Chemistry for Sanitary Engineers." McGraw-Hill, New York, 1967.
9. Stage digestion of waste water sludge, Water Control Federation.
10. C. E. Woods and J. F. Malina, Jr., Development of the anaerobic contact process, *J. Water Pollut. Control Fed.*
11. G. Schroepter and N. R. Ziemke, "Industrial Wastes."

Chapter VII

FUELS AND CHEMICALS FROM CROPS

Henry R. Bungay* and Roscoe F. Ward
DIVISION OF SOLAR ENERGY
ENERGY RESEARCH AND DEVELOPMENT ADMINISTRATION
WASHINGTON, D.C.

I. INTRODUCTION

Low-priced fuels and chemical feedstocks arc kcys to the good life. Cheap transportation lends great individual mobility to our citizens, allowing them to live where they like and to travel as it suits them. Traffic accidents, pollution, noise, and overcrowding are technical problems which can and must be alleviated, but freedom to travel has become a privilege

* Present address: Department of Chemical and Environmental Engineering, Rensselaer Polytechnic Institute, Troy, New York.

demanded by our people. Prohibitively expensive fuels could change our life-styles in a way that most of us would find seriously objectionable.

Vitality of our industries also depends on keeping products affordable, thus maintaining wide markets. Petrochemicals for solvents, plastics, and a myriad of other products used by industry or by private consumers or as intermediates to make other products must be inexpensive or else high costs will reduce sales, leading to fewer jobs and a lower standard of living.

As petroleum and natural gas resources dwindle and prices rise, other sources of fuels and chemical feedstocks become important. Coal can furnish most of those compounds now classed as petrochemicals but at increased costs. Ultimately, however, renewable resources must dominate over nonrenewable resources whose costs can only climb as it becomes more difficult to drill or dig deeper to get them.

This discussion assumes that in the long run fuels will not be important for heating, cooling, or stationary power needs. Other means for solar energy conversion have great promise in fixed installations for public buildings, factories, and homes. Direct conversion of sunlight to energy should be more efficient than growing biomass which must be converted to fuels which lead to energy. As these direct conversion technologies reach fruition, lower needs for petroleum will prolong its economic life. Nevertheless, biomass will someday become the most economic source of fuels and organic chemicals. Whereas other solar energy technologies are faced with difficulties in storage, biomass and its derived fuels are easily stored.

Green plants already serve as commercially important sources of chemicals, mostly carbohydrates or compounds derived from carbohydrates by fermentation or by chemical processing. Other compounds formerly produced by fermentation of carbohydrates are currently derived from petroleum because overall costs are lower. As economic factors shift, it makes a great deal of sense to reexamine old fermentation processes and to develop new technologies for biomass conversion. Not only can there be direct benefits to the United States but other nations will follow our leads.

II. BIOMASS SOURCES

Other chapters deal heavily with wastes; however, the need for fuels is far too great to be satisfied using just municipal and agricultural wastes as starting materials. Even the animal and silvicultural residues currently produced fall far short of U.S. energy needs. Logically, the accumulation of vast amounts of both waste and biomass grown for energy should be given prime attention. This chapter focuses on the latter. Waste accumulation is considered in Chapter I.

It is essential to identify the most promising sources of biomass and to select appropriate conversion processes. To get massive amounts, crops will be grown especially either partially or totally for conversion to fuels and chemicals. Even if other solar technologies satisfy the energy needs of fixed sites, fuels for highway and farm vehicles, aircraft, boats, and for supplementation of solar installations must come from biomass deriving solar energy from enormous areas exposed to sunlight. These areas should not have more valuable uses such as conventional agriculture or be recreational parks or nature reserves. Regional restrictions on land use can rule out some locations. The way to minimize the area required is to maximize the crop yield per unit area. Yields depend on geography, with warm, wet locations having the greatest potential productivity. High yields will require fertilization and intensive management.

A. Aquatic Biomass

Aquatic plant growth has different problems from those for terrestrial plants. Algae and their relatives such as kelp have been grown at very high rates in experimental units. Although these rates are unlikely to be reached in the field, expected productivities are much higher than those of most other plants. Table I shows some comparisons. Freshwater plants such as water hyacinth and duckweed are also possible candidates for energy crops, but public lakes and streams could be covered with plant growth only after overthrowing our democratic processes. Constructing impoundments of water with sufficient area for growing energy crops seems prohibitively expensive. Salt marshes or coastal lagoons could furnish inexpensive areas, but there may not be enough to grow large amounts of energy crops.

The open oceans represent the greatest underutilized areas of the earth's surface. These waters are low in fertility and support relatively little plant growth. Bottom sediments are rich in nutrients; upwelling of bottom waters results in abundant growth as exemplified by the anchovy industry off the coasts of Peru. Projects with the U.S. Naval Underseas Center and California Institute of Technology have investigated growing the giant kelp *Microcystis* on upwelled ocean waters. Kelp is an attached plant, and the proposed kelp farm will fasten seedlings to a net below the water surface. Periodically, ships with cutters will harvest a portion of the adult growth and transport it to shore for processing. Preliminary cost estimates are not very attractive, and the engineering problems are frightening for kelp farming.

Microscopic algae are costly to harvest because they filter poorly and centrifugation is much too expensive. It is anticipated that the collection problem can be solved. If not, larger algae grow less rapidly but are readily concentrated by screening or sedimentation. Sedimentation, however, is a highly undesirable feature for algae that are to be grown in deep waters because they sink from the photic zone. Either mechanical supports or selection of species that float would be required for algal forms in the oceans.

B. Terrestrial Plants

Some of the crops listed in Table I could provide both food and energy. For example, ears of corn would supply protein and starch while wastes and stalks could be converted to fuels and chemicals. At present, about 7% of corn is intended for human use and most of the rest is fed to cattle. Cattle silage is the whole plant except for the roots. By changing our diets to less meat, existing corn could serve as a food and energy crop.

Following this reasoning further, any plant material contains protein, and most proteins are good human nutrients. Ideally each possible product from a resource should be considered on its own economic merits, and fuels could be the main products or could be by-products. Several of the biomass conversion processes could accept a variety of organic leftovers as supplements for making methane or other chemicals.

This line of reasoning raises the status of the portion of crops which will go to producing energy. Rather than treating this nonfood element as a waste, or even as a by-product, it treats such matter as a coproduct. In utilizing this approach, the problem raised by the Office of Technology Assessment of the U.S. Congress—that we could be criticized for growing energy in a food-short world—would be answered.

Another way of looking at terrestrial biomass is in terms of its composition. Some crops are high in carbohydrates: Sugarcane, sugar beets, and sweet sorghum are rich in sucrose while corn and potatoes store starch. Stems and branches derive their structural strength from cellulose which is a polymer of glucose. Unfortunately, cellulose is less readily hydrolyzed than is starch, and in most biomass cellulose occurs in a matrix of lignin and hemicelluloses which seriously interferes with the action of cellulose enzymes. Although cellulose is the most common polysaccharide, it cannot be used directly in any of the existing commercial fermentation processses.

Microorganisms have storage and structural compounds that are different than those of higher plants. Laminarin and chittin are typical structural materials, and polyhydroxybutyrate and lipids are common storage compounds in microorganisms. Little consideration has yet been given to microbial cells as the major nutrient in any fermentation medium other than sewage sludge, which is very high in aerobic microorganisms from previous sewage treatment steps.

The point of this discussion is that for some conversion processes, the

TABLE I

Productivity of Agricultural Crops, Forest, and Seaweeds on an Annual Basis

Climate	Area	Yield[a]
Temperate		
Rye grass	U.K	23
Kale	U.K.	21
Sorghum	U.S., Illinois	16
Maize	Japan	26
	Iowa	16
	U.S., Kentucky	22
Potatoes	The Netherlands	22
Sugar beets	U.S., Washington	32
Wheat (spring)	U.S., Washington	30
Sycamore	U.S. (S.E.–N.C.)	12
Loblolly pine	U.S. (S.E.)	10
Hybrid popular	U.S. (N.C.)	20
Aspen	U.S. (N.E.)	4
Seaweeds	U.S., Massachusetts	33
Subtropical		
Alfalfa	U.S., California	33
Sorghum	U.S., California	47
Bermuda grass	U.S., Georgia	27
Sugar beets	U.S., California	42
Wheat	Mexico	18
Rice	U.S., California	22
Eucalyptus	U.S., California, Louisiana, Florida	26
Seaweeds	U.S., Florida	73
Tropical		
Napier Grass	Puerto Rico	85
Sugar cane	Hawaii	64
Sugar beets	Hawaii (2 crops)	31
Maize	Peru	26
Rice	Peru	22

[a] Tons of total dry matter in metric tons/hectare/year.

composition of biomass is of minor concern, but for the fermentation processes it is an overriding matter. Two other factors are moisture content and ease of subdivision. Vaporizing the water in wet biomass wastes energy from thermal processing; thus, relatively dry biomass is essential. All conversion processes, even direct burning, require subdivision. If a process needs finely divided biomass as feed, grinding costs could be excessive for a material such as wood. A total systems approach must be used in choosing the type of biomass, growing sites, and conversion processes.

III. ESTABLISHING THE ENERGY FARM

A. Operational Considerations

1. Unit Operations of an Energy Farm Different types of biomass can bypass certain operations, but an energy farm might require the following operations: (1) site preparation, (2) sowing or planting, (3) cultivation, (4) cutting or harvesting, (5) collection and transportation, (6) shredding or grinding, (7) conversion, and (8) packaging and distribution of product. These are shown in Fig. 1.

If the crop were corn, all these operations would be required; some would be seasonal but repeated several times. Crops that copice can avoid operations 1 and 2 for a number of years. Eucalyptus trees, for example, grow new stalks from old roots, but after four crops vigor declines, necessitating pulling stumps and replanting. Labor costs must be minimized by mechanizing and automating as much as possible.

Some crops eliminate some operations but are costly for others. Algae, for example, may have negligible charges for seeding and require no subdivision but have quite expensive capital requirements for leveling, pond construction, pumps, and the like. Cultivation can be costly if supplemental nutrients or carbon dioxide are needed.

2. Growing of Biomass Plants get carbon dioxide from air, and some can fix atmospheric nitrogen. However, most need to be supplied with nitrogen compounds as well as phosphates and trace minerals. Thermal processing of biomass to produce fuels will produce ash or charcoal ulti-

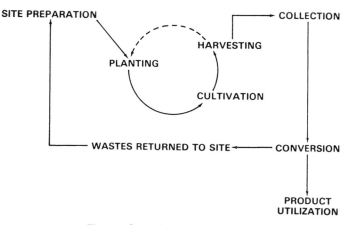

Fig. 1.   Operations at an energy farm.

mately becoming ash, and this ash is worth returning to the growth sites as a source of minerals. Nitrogen compounds may go out of the stock to give ash of low nitrogen content. Replenishing nitrogen in the soil may be a major expense.

Waste streams from bioprocessing should be rich in fixed nitrogen. Breakdown of biochemical nitrogen usually yields ammonia or water-soluble compounds that are in the liquid wastes. This can be recycled to the growth area for its water, minerals, and fixed nitrogen.

3. Objections to Farming for Fuel Crops   In order to achieve the high productivities of biomass required for energy farming to be profitable, the entire plant above the ground would be collected. This differs greatly from current practices of both forestry and food cropping because trimmings and residues are left to rot. Some enrichment of the soil results. However, erosion of the land by water or wind is less when protected by residues, and federal and state directives can insist on specific materials being left on the site. Alternatives such as terracing to prevent erosion by water or constructing windbreaks are expensive. Although not insurmountable, erosion problems are serious and can impact unfavorably on the economics of biomass production.

The concept of biomass as a renewable resource has been challenged. Farming, as we know it today, is not an operation of merely taking from the land; large amounts of fertilizer are added. Trace constituents of fertilizer are of little consequence, but nitrogen and phosphorus could become overly expensive. Fixing nitrogen chemically is energy intensive, thus it grows more expensive each year. Biological fixation of nitrogen by microorganisms in the root modules of legumes or by other microorganisms will someday be controlled effectively. In the meantime, current means of replenishing nitrogen in the soil will suffice for establishing energy farming. The key nonrenewable ingredient in fertilizer is phosphate. Phosphate rock deposits should be mined out in 400 to 500 years, which is a shorter life span than that predicted for coal mines. Phosphate availability thus can be a very real restraint on obtaining biomass.

Recycling of nitrogen or phosphorus to the soil with wastes from the bio-mass-to-fuel conversion process must be considered. Ash from a thermal conversion process could present difficulties if the needed elements are in chemical forms with poor solubility. This is not likely to be of much importance compared to the inefficiency of application whereby much of the fertilizer is washed off or is percolated through the soil. Recycle will not be very helpful if most of it is gone before being incorporated into the plants. Marine farming has the major plus of using ocean waters that can be fertilized from land runoff or from bottom sediments.

B.   Energy Farm Products

To have an impact on petroleum requirements, anything made from biomass should have a potential market of at least several million pounds per
year. A fuel to substitute completely for petroleum should be produced at
roughly 15 million barrels per day, but only 6% of petroleum goes to
petrochemicals. If other solar technologies are ready before petroleum
reserves are hopelessly low, petroleum prices will drop drastically as
production capacity overwhelms need. This could mean that scenarios
based on continued inflation of petroleum prices could be wrong if demand
drops so much that OPEC solidarity wavers. Reasoning this way, chemicals
from biomass should be competitive with petrochemicals in terms of
technology and should not depend on oil producers foolishly creating a
price structure in which their products are noncompetitive.

A number of volume chemicals come from petroleum or natural gas.
Some have large sales volume mainly because they are cheap; thus different
compounds with the same uses could displace them if priced lower. Some of
the most important bulk chemicals are now discussed briefly.

1.   Synthesis Gas   The combination of hydrogen and carbon monoxide in
various proportions is termed synthesis gas. Alternate routes for its production are

(1) $C + H_2O = CO + H_2,$
(2) $CH_4 + H_2O = CO + 3H_2,$
(3) $CH_4 + \frac{1}{2}O_2 = CO + 2H_2.$

It should be possible to convert biomass to synthesis gas fairly easily and
thus to spare natural gas and petroleum. However, synthesis gas is very
cheap, and thus it may be more profitable to make something else. It is better suited to pipelines than to the propulsion of vehicles. However, it is a
possible starting point for synthesizing many different organic molecules.
Synthesis gas is considered in Chapters II and IX.

2.   Ammonia   Second only to sulfuric acid in tons produced, this
chemical is approaching sales of 20 million tons/year. Since it does not
contain carbon, it might be overlooked as being related to petroleum.
Ammonia is made by the Haber process from nitrogen and hydrogen and
requires 545 kw to obtain 1 ton. Petroleum or natural gas thus can be a
source of hydrogen for ammonia and may drive the generators to produce
the required electricity.

Ammonia has had very little attention as a product from biomass. All living matter contains proteins whose amide linkages are broken during degradation to yield amino groups. Further degradation can be deamination to

ammonia which will be soluble ammonium ion except at high pH. Recovery by an inexpensive method could make ammonia from biomass an attractive product with significant impact on the U.S. energy picture.

3. Methanol   Methanol uses natural gas as the usual starting material for its production. Thermal processing of biomass often yields methanol. This fuel is discussed in Chapter XI.

4. Formaldehyde   This organic chemical is usually produced cheaply from methanol. It is probably not a likely product from any bioconversion.

5. Oxo Alcohols   Primary alkenes react with carbon monoxide and hydrogen to give alcohols such as *n*-butanol, 2-ethyl hexanol, isobutyl alcohol, isooctyl alcohol, and decanol. The market for the first two on the list approaches half a million tons of each while the others are about one-fourth of that. Several alcohols were made by fermentation for years before petroleum became a more economical starting material. Reexamination of these processes after being almost ignored for 20 years could well show biomass to be an economically viable source.

6. Aromatics   Crude oil contains about 1% benzene or cycloparaffins easily converted to benzene. Over 2% is toluene or methyl cycloparaffins. If benzene is recovered, by-product toluene is available far in excess of market needs. Similarly, xylenes as by-products from benzene production would be plentiful. Biomass as a source of aromatics would seem to hinge on lignin which is a highly cross-linked polymer of hydroxy- and methoxy-substituted aromatic compounds. Biological degradation of lignin oxidizes and opens the rings before separating them, thus few aromatic products are likely to be formed. This is to be expected because common mechanisms for detoxification oxidize the rings, and aromatic compounds are not tolerable in most biological systems. Thus thermal processing of lignin may produce aromatics, but bioconversion is not a likely approach. Heavy liquids are discussed in Chapter VIII.

7. Ethylene   This very important petrochemical can be obtained from biomass in some thermal conversion processes but it is not a probable product of bioconversion. Although its sales are in the range of 15 million tons/year, ethylene is very cheap. It makes more sense to convert biomass to something usually made from ethylene rather than to strive for this very inexpensive starting material. It also is unlikely to be economical to convert higher value chemicals such as ethanol to ethylene unless markets for ethylene are overwhelming.

8. Heterocyclics   Only furfural or its derivatives are likely heterocyclic products from biomass. Thermal processing of wood gives appreciable yields of such heterocyclics, and markets would depend on costs. Plastics based on furfural are fairly easily interchangeable with other plastics. If produced very cheaply, furfural plastics could compete very favorably with those made from petrochemicals. Phenolic compounds of furfural using phenols from the lignin component of wood deserve consideration as high-sales products.

Biochemical reactions of carbohydrates could give rise to furfural and hydroxymethyl furfural; a bioconversion is thus deserving of consideration. Currently, however, there is little known about biological production of furfurals.

9. Gum Naval Stores The exudates from coniferous trees contain organic chemicals which are used as solvents, chemical intermediates, additives for paper or fabrics, and for many other purposes. Best known of the pine products are turpentine, which is a mixture of liquid terpenes, and rosin, which is high in carboxylic acid derivatives of aromatic ring compounds such as phenanthrene. The naval stores industry has a competitive disadvantage because yields per tree are only a few pounds per year and manual labor is required to slash a tree, to apply stimulants for exudation, to fasten a collector which must be relocated often, and to gather the collected gum. Furthermore, only mature trees about 25 years old are ready for tapping and produce gum for about 10 years.

Turpentine production is roughly 30 million gal/year mostly for solvent purposes, but its main constituents, α- and β-pinene and other terpenes in lesser amounts, have good potential as chemical synthesis feedstocks. Pine oil and synthetic pine oil made from pinenes are widely used in the textile industry for wet processing, in sanitary chemical products for pleasant odors and cleaning properties, and for some solvent applications. Several perfume chemicals can be synthesized from the ingredients of turpentine. The uses and value of turpentine chemicals could grow rapidly if there were abundant supplies.

The resinous solid remaining after distillation of the turpentine fraction is known as gum rosin. It is excellent for sizing paper and textiles. While generally equivalent in applications to petroleum rosins, gum rosin is higher priced and its annual production is declining. As with turpentine, the ingredients of gum rosin could be quite valuable as chemical intermediates.

A recent change in the gum naval stores industry is the collection of significant amounts of pine chemicals as by-products of pulping. There are severe losses of volatile compounds as cut pine trees wait to be processed. Vapors from hot, caustic pulping can be condensed to yield turpentine of slightly lesser quality than that obtained from tapping trees. In the next step of pulping, there is a layer formed on the cooler liquid of a "soap" resulting

from hydrolysis of pine gums. These are sodium salts of carboxylic derivatives of polynuclear hydrocarbons. Although considered quite inferior to
natural resin for resin purposes, there might be a number of uses as
chemical feedstocks if available in large, dependable amounts. Growth of
coniferous trees for energy could be compatible with a large industry for by-
product recovery but not with conventional gum naval stores. Harvesting of
immature trees too young for tapping rules out old methods, but the entire
tree, needles and all, contains appreciable pine chemicals. Either thermal
processing, extraction, or bioconversion could provide important products
from a major new industry.

10. Cellulose Derivatives Cellulose acetate, carboxymethylcellulose,
plastics fabricated from cellulose, and other cellulose derivatives are well
established commercially. Research on producing fuels and chemicals from
biomass might well uncover ways to produce high grades of cellulose suitable as feedstocks for existing factories. If a small fraction of the biomass
produced for energy were diverted to cellulose derivatives, the economic
credits would be significant.

C. Conversion Processes

1. Methane by Anaerobic Digestion of Organic Matter All biomasses
seem to be acceptable for anaerobic digestion. Proof of digestibility
appeared in the research literature long ago for some, while others such as
kelp were recently shown to digest well. Research and development are
needed for anaerobic digestion to optimize yields and to speed the reactions
to achieve shorter detention times which allow smaller vessels and lower
equipment costs. Some current research topics are pretreatment of biomass
to increase digestibility and optimization of process parameters to improve
performance.

Cost estimates indicate that anaerobic production of methane should be
economic when collection of biomass is not too expensive. Pilot scale
anaerobic digesters located on cattle feedlots or other places with abundant
wastes are learning about possible problems and providing data for a defini-
tive cost analysis of this system. If successful, industry could be persuaded
to invest in large factories to supply methane for fuel gas. This approach is
explored in Chapter VI. Figure 2 shows a general flowsheet for the production of methane by anaerobic digestion.

2. Fermentation of Biomass The fibrous components of many forms of
biomass contain cellulose, almost always in a matrix of hemicelluloses and
lignin. Cellulose can be split by chemical hydrolysis, by enzymes, or by
chemical treatment followed by enzymatic hydrolysis to glucose which is a

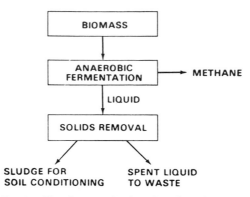

Fig. 2. Flowsheet: production of methane from biomass.

well-established nutrient for fermentations. Chemical hydrolysis conditions destroy glucose, so overtreatment must be avoided. Fermentation of glucose can be directed toward ethanol, acetone, butanol, and a number of other organic chemicals which are equivalent to or can substitute for compounds currently obtained from petroleum. In other words, cellulose can be a source of cheap glucose which can be the basis of economic fermentations for products which are fuels or chemical feedstocks.

The feasibility of producing fuels and other organic chemicals from cellulose is well proven, but profitability is marginal. Pretreatments to reduce particle size, to remove lignin which impedes cellulose hydrolysis, and to render the cellulose more amenable to hydrolysis are costly. Ethanol from cellulose requires four main steps: (1) separate fermentation to produce the enzyme cellulase, (2) recovery of the enzyme, (3) enzymatic hydrolysis of cellulosic biomass, and (4) fermentation to produce ethanol. Several groups are performing process development. Some of this will have to be performed on equipment of pilot size to get accurate costs and to learn about scale-up.

3. *Hydrogen by Biophotolysis* Hydrogen is a product of several biochemical reactions. During photosynthesis, oxygen is liberated in gaseous form while ordinarily hydrogen is incorporated into biochemicals. The photons which strike the photosynthetic pigments supply energy which is used to split water into hydrogen and oxygen, and the hydrogen can be liberated from its acceptor compounds by hydrogenase enzymes. Photosynthesis accompanied by release of hydrogen has been demonstrated in the laboratory by a variety of methods, but none of them is yet commercially practical. A major difficulty at present is maintaining reducing conditions so that liberated oxygen does not inhibit hydrogenase action.

4. Pyrolysis of Biomass   Pyrolysis or its old name "destructive distillation of wood" has as its products water, organic compounds, gases, tars,
and char. Among the gases are hydrogen and methane, which have fuel
value, and several of the organic compounds are commonly regarded as
fuels or petrochemicals. This process could well be the best way to convert
biomass to chemicals, but no discussion is given here because of the detailed
coverage in other chapters of this book.

5. Gasification of Biomass   Somewhat similar to pyrolysis but markedly
different in the product mix is gasification. Hydrogen or carbon monoxide,
or both, depending on the particular gasification process, provides reductive
conditions when biomass is heated so that oils are predominant products.
Less char and fewer simple organic compounds are formed, and the oils can
be substituted for common fuel oil. Again details are presented in other
chapters.

D.   Disposal of Energy Farm Wastes

1. Anaerobic Digestion   Several authors have fostered the concept that
biomass can be digested anaerobically to methane and carbon dioxide with
the leftover sludge having appreciable value as fertilizer. This is something
for nothing: biomass in − fuel + upgraded biomass out. Years of experience with domestic sewage treatment and with treatment of industrial
wastes have shown that waste sludge after anaerobic digestion is not a very
good fertilizer. These digesters are fed the sediments from the plant inlet
streams plus the settled microbial growth obtained by aerobic treatment of
the soluble portion of the input streams. As biomass goes, this is rich
material. Nevertheless, digestion of this rich sludge reduces its solids
volume, gives methane, and produces a waste sludge that is low in nutrients.
A crude picture of digestion is that of a succession of organisms growing,
dying, and serving as the feed for the next generation. After a long
residence, most of the organic nutrient value has been expended with the
inefficiencies of the growth stages. The waste sludge is not an asset because
collection and drying costs must be factored into its sale price. Often, it is
given to those willing to carry it away for use as a soil conditioner.
   Digestion of plant biomass may differ greatly from digestion of sewage
sludge. Verification of fertilizer value of the waste sludge is necessary before
hopes are raised too high for an economic credit.

2. Fermentation   Industrial fermentation wastes constitute a disposal
problem in that their biological oxygen demand (BOD) is high. Treatment is
usually easy but costly because biological decomposition is rapid only if there

are high degrees of aeration and mixing. Fermentation of sugars from bio-mass should be very similar to a number of industrial fermentations.

Recycling of fermentation wastes to the growing areas is an attractive alternative to disposal. Nuisances such as odor or unsightly appearance are not likely. Nutrient value should be high, and desirable minerals will be present.

3.   Cellulose Pretreatment and Hydrolysis   Hot acid or alkaline pretreat-ment of cellulose will furnish a miserable waste. In the large amounts generated at an energy factory, there will be major expenses for neutraliza-tion and disposal. Lignin and other organic material will result in an unpleasant and messy solution. The value of some ingredients as fertilizer will be offset by having too much salt which will have only small markets if crystallized for sale. In fact, neutralization has so many bad features, that recovery of the acid or caustic seems essential. This will be expensive, but waste streams from the recovery operation may be acceptable for recycle to the growing areas.

Side streams from cellulose hydrolysis will contain undissolved solids which can be recycled to the pretreatment step. Buildup of recalcitrant material resistant to hydrolysis will necessitate bleed-off of a portion. This may have some fertilizer value and should not be harmful to growth.

4.   Thermal Processing   Most of the fuel fractions will leave the factory and thus disappear as factors in waste treatment. If char is burned at the factory, ash can probably be used as a mineral supplement for growth. If sold for burning remote from the farm, it will be costly to return the ash.

IV.   IMPLICATIONS OF ENERGY FARMING

Fuels from biomass will impact on international relationships and on society in a number of significant ways. Only those that are peculiar to energy farming are covered here even though this chapter probably takes a broader view than those chapters considering only wastes. Needless to say, another major source of fuels will disrupt prices and markets for coal and oil, create new jobs, cause old jobs to disappear, shift the economics of some states up while others drop, and probably lead to a plethora of new laws designed to insulate special interest groups from the painful gyrations of a free enterprise system. Until something is known about what the bio-mass will be, what conversion technology is appropriate, and when it will be ready, any economic projections have little meaning. However, it may be useful to generalize on some of the most probable consequences of energy farming.

One very important fact is that energy farms will be highly visible. The efficiency of photosynthesis is reasonably good, but the national energy requirements equate to vast areas exposed to sunlight. Those who promote fuels from biomass tend to state or to imply that land of little value will be used for energy farms. While crops for energy may have somewhat more flexibility for suitable soil, for climate, and for water than food crops, it will almost always be more productive to grow them on fertile soil and with favorable conditions. Aquatic plants avoid land requirements but present problems in securing nutrients. The open oceans are not very fertile; cultivating plants in good yields will require artificial upwelling of sediments from the bottom or surface application of fertilizers. Coastal waters, salt marshes, inland swamps, or artificial impoundments may not be cheap areas when preparation costs are considered. Really cheap land in this country is desert, mountainous, or remote. Harvesting costs can be prohibitive if land is rocky or inclined too steeply for easy traverse by machinery. If too remote, transporting the crops or the fuels produced from them will be very expensive. This means that energy farms will probably not be hidden away from view but will take over existing cultivated lands or be fairly near them.

Energy farms will be good neighbors. Photosynthesis is mute and most plants have pleasant odors. Release of oxygen and removal of carbon dioxide from air are to be desired. The unpleasantness will come during sowing or harvest if noisy, polluting equipment is used. Even so, these operations will be periodic and oftentime will be in the interior of the farm far from towns or dwellings.

The appearance of an energy farm will not be bad. If the crop is similar to conventional food crops, there will be nearly bare fields after harvest and occasional plowing or clearing. With good perennials, years may pass before the land requires tillage. Applications of fertilizers, pesticides, herbicides, growth promoters, and the like are well-accepted operations that should cause little outcry. If the preferred energy crop is trees, the harvesting time will probably be 6 to 10 years. Right after harvest, the land will appear blighted. The next year will see new growth, and at any time 80 to 90% of the farm will be verdant.

There seems little question that the growing areas of the energy farms will be aesthetically acceptable, so attention can be turned to the factories for converting biomass to fuels. Thermal conversion plants will have characteristics similar to those of a petroleum refinery with tanks, stacks, and fractionation towers. Although not the cleanest of factories, air and water pollution should not be unreasonable. Odors should not be problems, and the wastewater should be quite acceptable for irrigation of the growing areas.

Aquatic farming will be seen by few people. However, there could be major consequences for the environment. Massive upwelling of bottom waters will lower the surface temperature of sufficient ocean areas to affect local climates. Lowered rainfall on adjacent lands could result in a large net financial loss if more valuable crops suffer as energy crops prosper. A long-term effect could be hastening of the next ice age. Aquatic crops grown on artificial ponds or impoundments on shore could result in more humid conditions in surrounding areas. Daily temperature fluctuations would be damped somewhat. It is difficult to predict whether these climatic effects would be favorable, but agriculture may benefit while the humidity causes personal discomfort.

V. CONCLUSION

Fuels and petrochemical substitutes can be obtained from renewable biomass now, but the economics are marginal or unfavorable. To supply a significant fraction of U.S. energy consumption, enormous amounts of biomass are needed. This presents great engineering challenges for site preparation, cultivation, harvesting, and transportation to factories for conversion to fuels. A high degree of cost consciousness is needed as these operations are developed, and the factories will be so large that design economics are essential. Nevertheless, fuels and chemicals from biomass will be a major industry. By using our brains and applying our talents, this can be one of the best approaches to alleviating the energy crisis.

Chapter VIII

THE PRODUCTION OF OIL FROM WOOD WASTE

Herbert R. Appell
PITTSBURGH ENERGY RESEARCH CENTER
ENERGY RESEARCH AND DEVELOPMENT ADMINISTRATION
PITTSBURGH, PENNSYLVANIA

I. INTRODUCTION

Cellulose has been converted to liquid materials by high pressure hydrogenation [9] and to a bitumenlike material by treatment with caustic at elevated temperatures and pressures. This latter work was initiated by Berl

121

[6,7] and continued by Heinemann [11–13] who showed that bitumen could be hydrogenated to petroleumlike materials.

The use of carbon monoxide in converting carbohydrates to oil stems from the early work of Fischer and Schrader [10] who demonstrated the effectiveness of carbon monoxide in solubilizing brown coal. A reinvestigation of this approach to coal liquefaction at the Pittsburgh Energy Research Center showed that the chemistry of the process could be applied to carbohydrates [1]. This observation led to a detailed investigation of the scope of the reaction with carbohydrates and an investigation of the mechanism of hydrogenation with carbon monoxide [2–4]. Inasmuch as wood wastes consist predominantly of carbohydrates, the technology of hydrogenation using carbon monoxide was readily adaptable to converting wood to oil. The oil obtained from wood is a viscous material and in some cases may be classified as a bitumen (see Section VI).

A. Experimental Procedures

Most of the laboratory-scale experimental work was conducted in a 500-ml stainless steel autoclave. Heat and agitation were supplied by a rocking furnace. The feedstock, water, and catalyst were charged to the cold autoclave, carbon monoxide was added to the desired pressure, and the autoclave was then brought to operating temperature. The reaction time reported does not include the heating and cooling periods. These periods, where significant reaction may have occurred, are about 1 hr for runs at 250°C and about 2 hr for runs at 350°C and above.

The reaction product was flushed from the autoclave with solvent, and the product oil was extracted in a Soxhlet unit. Acetone was used as an extraction solvent for product from runs conducted at 300°C or less, and benzene was used for product from runs above 300°C. The oil, or bitumen, was recovered by flashing off the solvent; a rotary evaporator heated in a hot water bath was the preferred method of removing solvent. The very low content of low boiling materials in the product made it possible to get a good separation of solvent and product. The percent conversion is determined by subtracting the percentage of insoluble organic residue, remaining after solvent extraction, from 100. All calculations are on a moisture- and ash-free (maf) basis.

Some work was also conducted in 1- and 5-liter autoclaves when the purpose was to obtain enough product for supplying samples for evaluation and further investigation. A 500-ml magnetically stirred stainless steel autoclave was used for determining how product quality was affected by the rate of heating the reaction mixture to operating temperature. This stirred autoclave could be brought up to 350°C from 200°C, below which very little reaction occurs, in 15 to 20 min and cooled in less than 1 min by means of an internal cooling coil.

II. REACTANTS

A. Substrates

Water-soluble carbohydrates and carbohydrates that can be hydrolyzed under alkaline conditions can be converted to a bitumen or oil at temperatures of 250 to 400°C. Materials which have been demonstrated to undergo this conversion include glucose, lactose, sucrose, cornstalks and leaves, newsprint, hard- and softwood sawdust, bark, pine needles and twigs, bovine manure, and sewage sludge. This reaction has also been applied to low-rank coals at temperatures of 380 to 450°C [4].

Although a wide variety of carbohydrate-type materials can be utilized, some are more readily converted than others. In general, more carbon monoxide is consumed as the feedstock becomes more difficult to liquefy. Not only is more carbon monoxide consumed in hydrogenating the more refractory forms of carbohydrates, but more carbon monoxide is consumed by the water–gas shift reaction which occurs readily at the operating conditions of this process (see Section VII).

Although this chapter deals primarily with wood, the technology appears applicable to all the carbohydrate-type wastes evaluated. Lignin, a major component of many cellulosic materials, is also converted to an oil at all but the mildest processing conditions used for cellulose and does not appear to impose any processing problems.

In general, wastes with a high lignin content or those that consist largely of high-molecular-weight, crystalline cellulose are more difficult to convert than other wastes. Douglas fir bark is an example of a waste that is difficult to convert. This bark has a very high lignin content and contains considerable tannic acids. Specifically, Table I shows a considerable spread of carbon monoxide consumption values per 100 gm of oil formed for five typical cellulosic wastes. The figure for corncobs is low because a low ratio of carbon monoxide to corncobs was used (corncobs react readily because of the presence of considerable hemicellulose, which hydrolyzes and breaks down readily). The figure for pine bark is high because this material is less reactive, and less oil was formed in the hour at process conditions than with the other wastes. Sodium bicarbonate was used as a catalyst in this set of experiments (see also Sections III.C and VII).

B. Water

Water is a necessary component of the mixture undergoing the oil-forming reaction. Water not only supplies the hydrogen for the reaction, but is necessary for the hydrolysis of the high-molecular-weight carbohydrates present. In addition, it acts as a solvent and vehicle for the reaction and also

TABLE I

Conversion of Various Cellulosic Wastes to Oil[a]

Raw material	Amount (gm)	Time at reaction temp. (hr)	Maximum pressure (psig)	Conversion (%)	Oil yield (%)	CO used (gm/100 gm oil)
Cornstalks	50	1	1760	95	42	21
Corncobs	100	1	2400	95	38	9.7
Rice hulls	50	1	1600	97	40	25
Newsprint	50	0.5	1640	88	40	21
Pine bark	50	1	1600	83	32	45

[a] 200 ml water, 10 gm NaHCO$_3$, 500 psig initial CO pressure, 250°C reaction temperature.

decreases the extent of polymerization of some of the highly reactive water-soluble intermediates.

There are disadvantages of water in the system: (1) The high partial pressure of steam raises the operating pressure to levels where capital costs would be high; (2) the heat required to bring water to the operating temperature and pressure adds considerably to the operating costs; and (3) the separation of the oil and water phases during the product recovery step is sometimes encumbered by emulsions.

Because of the high pressure developed, it does not appear practical to use water as a solvent at temperatures above 350°C. Most of the work using water was conducted at 250 to 325°C. The aqueous phase contains a significant content of reactive compounds, resulting from the breakdown of the carbohydrates. Upon standing in an open beaker, the aqueous product turns dark and begins to deposit a dark oily material within 24 hr. This is probably a result of the polymerization of compounds such as pyruvaldehyde and its analogs [14]. These reactive compounds are intermediates in the conversion of carbohydrates to oil and may be utilized by recycling the aqueous effluent.

Table II shows the results of eight successive autoclave runs on newsprint with recycling of the aqueous phase containing sodium bicarbonate as catalyst for the reaction. Makeup water was added to bring the volume of the catalyst solution to 200 ml in all runs. The oil yield on the first run is usually low because part of the cellulose hydrolysis products remain in the aqueous solution. Thereafter, an equilibrium concentration of soluble carbohydrates and carbohydrate derivatives is reached so that, in effect, all the hydrolyzed cellulose appears converted to bitumen, water, and carbon dioxide. After a few cycles, there was a decrease in conversion. This decrease in conversion and yield is most likely a consequence of the formation of organic acids which partially neutralize the alkaline catalyst and result in a slower rate of oil formation. The pH of the solution dropped to about 5 after the processing step. However, the conversion could be returned to the 90% level by increasing the reaction time. However, reduction of the carbon monoxide pressure caused a considerable drop in conversion.

The amount of soluble organics obtained is largest in the lower range of the effective operating temperatures (250–275°C) and when relatively large weight ratios of water to carbohydrate (4:1 or more) are used. Under these conditions, 10 to 15% of the original carbohydrate may remain as water-soluble organics. With recycle of the aqueous solution, these organics would eventually be converted to oil. Because of the accumulation of organic acids, it is expected that the aqueous phase would require regeneration via drying and combustion to remove the acids and thereby recover the sodium carbonate. In the case of high moisture feedstocks, more water is recovered than is required for recycle and some of the aqueous water must be removed from the system.

C. Reducing Gas

Carbon monoxide was used exclusively in the early experimental work at the Pittsburgh Energy Research Center. However, pure gas is not needed. Synthesis gas (a mix of hydrogen and carbon monoxide) or crude carbon monoxide containing considerable carbon dioxide can also be used. Inasmuch as carbon monoxide is much more reactive than hydrogen to liquefy-

TABLE II

Effect of Recycle Catalyst Solution on Cellulose Conversion[a]

Run	Initial CO pressure (psig)	Time (min)	Oil yield (%)	Conversion (%)
1	800	15	40	91
2	700	15	54	92
3	700	15	43	82
4	700	15	44	88
5	700	15	46	84
6	700	60	53	91
7	400	60	41	78
8	400	60	40	78

[a] 250°C, 50 gm newsprint, 10 gm $NaHCO_3$, 200 ml water. A small amount of water was added to the recycle solution after each run to keep the volume at 200 ml.

ing carbohydrates in the presence of water and the absence of active transition metal catalysts, a high carbon monoxide content is preferred.

The effect of carbon monoxide pressure on the conversion of newsprint to an acetone-soluble oil is shown in Table III. An excess of water was used in these experiments, and the differences in oil yield are believed to be entirely due to the carbon monoxide partial pressure. Although considerable oil was formed in the absence of carbon monoxide, the char formed at these conditions retained the shape of the initial charge of newsprint. At an initial pressure of 200 psig or more of carbon monoxide, the voluminous cellulose structure collapsed and only an oil layer and water layer were found at the end of the experiment.

Below 380°C, the rate of conversion of carbohydrates to oil or bitumen depends largely on the partial pressure of carbon monoxide and not the other gaseous constituents in the system. At temperatures of 300°C and less, hydrogen has very little effect on the reaction so long as water is present and transition metal catalysts are absent. At temperatures of 350 to 375°C hydrogen begins to have a small effect, and above 380°C the effect of hydrogen becomes more apparent in hydrocracking reactions and to a lesser extent in hydrogenation. Table IV shows that at 300°C hydrogen has little effect on acetophenone, whereas carbon monoxide produces a 43% yield of phenylethanol. As the temperature is increased to 350°C and then 380°C, more and more hydrocracking activity is observed in the presence of hydrogen, and yields of benzene and toluene increase. At all three temperatures, carbon monoxide is by far the more selective and active reducing agent.

Because of the low cost of synthesis gas, most of the recent work has been conducted using this gas in place of carbon monoxide. With synthesis

TABLE III

Effect of Carbon Monoxide Pressure on Cellulose Conversion[a]

Initial CO (psig)	Operating pressure (psig)	Oil yield (%)	Conversion (%)
0	960	24	78
100	1150	24	76
200	1380	32	83
300	1480	32	82
400	1500	34	84
500	1640	35	87
600	1840	40	90

[a] 50 gm newsprint, 200 ml water, 10 gm $NaHCO_3$, 1 hr at 250°C.

TABLE IV

Reduction of Acetophenone[a]

Temp. (°C)	Gas	H utilization[b] (%)	Product composition (%)[c]					
			$PhCOCH_3$	$PhCHOHCH_3$	PhH	PhMe	PhEt	$PhCHCH_2$
300	CO	77	56.7	43	—	—	—	0.2
	H_2		98.8	1.2	—	—	—	—
350	CO	62	55	40	0.5	0.2	1.6	1.8
	H_2		92	1.7	3	0.1	1.6	0.1
380	CO	53	54	27	3.2	1.8	6.0	2.0
	H_2		61	2.3	8.6	5.2	10.5	0.4

[a] 100 gm acetophenone, 25 gm water, 5 gm Na_2CO_3, 1000 psig, 1 hr.
[b] Hydrogen utilization: H utilized divided by H available from formate decomposition.
[c] Determined by GLC analysis.

gas, somewhat higher pressures are needed to achieve the same result. By using a starting pressure of 300 psig of carbon monoxide or a pressure of 500 psig of 1:1 synthesis gas, 98 to 99% conversion of the organic matter in wastes can be obtained. The corresponding operating pressure when using synthesis gas is near 2500 psig.

In processing lignite, higher temperatures (430–450°C) are used and both the hydrogen and the carbon monoxide are utilized [5]. The relatively large amounts of moisture present during carbohydrate processing limit the useful range of temperature to below 400°C because of the excessive pressures developed. Below 375°C molecular hydrogen takes little part in the reaction and behaves as an inert diluent.

It is possible to operate in the absence of carbon monoxide by substituting either formic acid or sodium formate as the reducing agent (see Section III). This substitution lowers the operating pressure at 250°C to 1000 to 1,100 psig. Table V shows that good conversion and oil yields were obtained in the presence of cresol as a solvent (see Section III.D) when using several formate combinations as long as a 2:1 ratio of water to wood was used. Decreasing the cresol-to-wood ratio from 1:1 to 0.5:1 did not appear to be detrimental. In this version of the process, the formates would have to be regenerated from the spent catalyst solution in a separate step. This technique would be less successful at temperatures above 250°C because the formates decompose too rapidly and would become depleted before the oil-forming reaction was complete. The major portion of the research in this report utilized carbon monoxide and higher temperatures and pressures because of improved product fluidity.

TABLE V

Liquefaction of Wood in the Absence of Carbon Monoxide[a]

Cresol solvent (gm)	Water (ml)	Catalyst		Pressure (psig)	Oil yield (%)	Conversion (%)
		Type	Amount (gm)			
50	100	HCO_2Na	1	1070	60	99.6
		HCO_2H	5			
50	100	HCO_2Na	5	1040	55	99.0
		HCO_2H	1			
50	100	HCO_2Na	2.5	1050	55	99.8
		HCO_2H	2.5			
50	50	HCO_2Na	2.5	1040	41	90
		HCO_2H	2.5			
25	100	HCO_2Na	2.5	1020	57	99.9
		HCO_2H	2.5			

[a] 50 gm softwood sawdust, 1 hr at 250°C.

III. OPERATING VARIABLES

A. Temperature

The conversion of carbohydrates to oil has been investigated over the temperature range 250 to 400°C. Below 250°C the reaction appears too slow to be of practical value. Temperature has little effect on conversion and oil yield in its effective range. There is a tendency for some of the product to carbonize at 400°C and above. At lower temperature the product contains considerable oxygen (near 20%) and this accounts for the relatively large yields.

If a solid or semisolid product (a bitumen) is acceptable, operation at temperatures near 250°C may be preferred because of the lower pressures and the lower carbon monoxide consumption.

The reaction temperature does have a significant influence on the viscosity and oxygen content of the product. The product obtained at 250°C is a soft, bitumenlike solid at room temperature, but becomes readily pourable as its temperature is raised to near 100°C. On the other hand, the oil formed at 380°C is a free-flowing liquid with a viscosity of 650 cs (centistokes) at 50°C and 102 cs at 88°C.

The transition of the physical state of water at its critical temperature (375°C) may complicate interpretation of results from work in the 350 to 400°C range. A moderate decrease in temperature may sometimes result in more of the reaction mixture existing as a liquid phase, and thus improve product quality. But when a major portion of the reactants is already in the liquid phase, a decrease in temperature can be expected to result in a product of higher viscosity and oxygen content.

A major advantage of low temperature (250°C) in batch tests is that the pressure developed by the steam and gases is usually about 1500 psig compared with almost 5000 psig at 400°C. Another is that very little carbon monoxide is consumed by the water–gas shift reaction at 250°C. As the temperature is raised, the liquefaction reaction proceeds well but the water–gas shift reaction begins to consume carbon monoxide, and a major advantage of low temperature operation begins to disappear.

In processing wood wastes, 325°C appears to be the lowest temperature at which a pourable product can be obtained (see Section III.D).

B. Pressure

The operating pressure is due to gases charged to the system, gases formed in the reaction, and steam formed by vaporization of the water charged to the reactor. Some water is formed by dehydration of carbohydrates, but carbon dioxide is the major gaseous product formed in the reaction. Carbon dioxide forms in two ways: (1) by the water–gas shift reaction and (2) by decomposition of cellulose or other waste material. Some hydrogen is also formed, most of it via the water–gas shift reaction. Small amounts of carbon monoxide, gaseous hydrocarbons, and other low boiling organic materials are formed by thermal decomposition of the organic matter, but their influence on the pressure is negligible.

If the temperature is above 374°C, the critical temperature of water, all the water is in the vapor phase, and the pressure due to water alone becomes as high as dictated by the quantity of water put into the autoclave. Below 375°C, if liquid water is present at reaction temperature, the vapor pressure of water is regulated by the solution concentration, thus making this pressure somewhat less than the known vapor pressure of pure water. If a relatively small amount of water is added initially to the autoclave, all water will vaporize, and the steam pressure at a given temperature will depend on the volume of the autoclave and the amount of water confined.

The effect of carbon monoxide pressure on oil yield and conversion was illustrated in Table III.

The operating pressures used in the research on the conversion of wood to oil covered the range 1000 to 5000 psig. The range of 1500 to 3500 appeared most useful with respect to product quality and carbon monoxide consumption.

C.   Catalysts

Some wastes do not require the addition of catalyst. The mineral content of wood, however, is low and not all of the proper type for catalytic activity, so the addition of catalyst is necessary.

Water-soluble alkaline compounds, such as sodium carbonate, are effective catalysts (Table VI); conversions of over 90% and oil yields of 45 to 50% are attainable with these catalysts. Generally, hydroxides, carbonates, bicarbonates, and formates of the alkali metal and alkaline earth groups are effective catalysts. At process conditions, these materials probably exist as a mixture of carbonates, bicarbonates, and formates as they undergo conversion from one form to the other.

The data in Table VI show that the alkali carbonates are more effective catalysts than stannous chloride or ferrous sulfate, which are typical catalysts used to hydrogenate coal. The ammonium cation (NH_4^+) is similar chemically to the alkali metal cations, and it is no surprise that ammonium hydroxide also gave high conversions of cellulose; but the oil formed in the presence of NH_4OH had a high nitrogen content (2–4%) and this is a disadvantage.

TABLE VI

Effect of Various Catalysts on Cellulose Conversion[a]

Catalyst	Water (gm)	Time (min)	Conversion (%)
		120	63
$FeSO_4$	1	120	69
K_2CO_3	0.2	120	80
Na_2CO_3	0.2	15	78
Na_2CO_3	0.2	120	81
Na_2CO_3	1	120	96
NH_4OH^b	1	120	73
NH_4OH^b	10	120	96
$SnCl_2$	1	120	77

[a] 20 gm filter paper, 40 ml water, 1500 psig initial CO pressure, 350°C. Operating pressure about 4800 psig.

[b] A 30% aqueous solution of NH_3.

The alkaline salts may have several effects:

(1) In the presence of carbon monoxide, they are converted to formates which are reducing agents and transfer hydrogen to the oxygenated or unsaturated compounds, and then are regenerated *in situ* (see Section VII). In other words, they serve as homogeneous catalysts for converting solid wastes to oil.

(2) Alkali carbonates are catalysts for the water–gas shift reaction.

(3) Alkaline materials are catalysts for many known organic rearrangements and disproportionations that yield materials containing less oxygen than the original carbohydrates.

(4) Alkaline salts may neutralize organic acids formed in the system. Without neutralization, these acids could promote charring.

Although originally alkaline, the pH of the recycle aqueous phase drops to about 5 during use, probably because small amounts of soluble organic acids form. The gradual decrease in conversion rate with reuse of catalyst solution suggests that acidic compounds are accumulating and that eventually regeneration of the solution may be necessary.

However, the most effective catalysts for converting cellulosic materials to oil at 350°C are high boiling heterocyclic nitrogen bases, such as isoquinoline, discussed in the following section. These dual-role catalyst vehicles or solvents are particularly effective in promoting the efficiency of hydrogen utilization at high temperatures (Table VII). Hydrogen utilization correlates well with conversion of carbohydrates to oil, and is a measure of

TABLE VII

Efficiency of Hydrogen Utilization[a]

Catalyst	Water (ml)	Conversion (%)	Final gas composition (vol %)			Hydrogen utilization (%)
			H_2	CO	CO_2	
None	40	73	18	47	33	45
Isoquinoline	20	98	6.8	49	41	83

[a] 40 gm soft pine, 1200 psig initial CO pressure, 380°C for 15 min.

catalyst effectiveness. Hydrogen utilization is defined arbitrarily as the hydrogen added to the cellulose divided by the hydrogen released by formate decomposition.

D. Solvents

At temperatures above 350°C, only a few solvents meet the requirements of stability, low vapor pressure, solvent action, and low to moderate cost. One of these solvents is anthracene oil, a by-product of coal-tar refining. Other excellent solvents are phenols and high boiling heterocyclic bases such as isoquinoline and alkylpyridines. These latter compounds are not only excellent solvents of high stability but also powerful catalysts for reactions leading to cellulose liquefaction. High conversions can be obtained by using smaller amounts of isoquinoline than of anthracene oil. Because of their cost, however, it will be necessary to recover the heterocyclic bases for recycling.

The effect of anthracene oil and isoquinoline in liquefying white pine wood chips at 380°C is shown in Table VIII. When either anthracene oil or

TABLE VIII

Effect of Solvent on Operating Pressure and Cellulose Conversion[a]

Water (ml)	Solvent (ml)		Time (min)	Operating pressure (psig)	Conversion (%)	Oil yield (%)
	Anthracene oil	Isoquinoline				
40			15	5850	73	22
20	40		15	4300	95	51
20	40		120	4300	96.5	53
20		5	15	4200	98	57

[a] 40 gm soft pine, 1200 psig initial CO pressure, 380°C.

isoquinoline replaced part of the water, the operating pressure was considerably lowered and conversions and oil yields increased. The isoquinoline was especially effective because it acted as a catalyst as well as a solvent.

The effectiveness of organic solvents is believed to be due in part to the extraction of reactive intermediates from the aqueous phase into the oil phase, where they are converted to oil. The high boiling solvent also provides a polar liquid medium that assists in the breakdown of the large carbohydrate molecules to soluble reactive species.

The objectives of the research on solvents were to investigate the use of recoverable solvents and to develop process conditions yielding an oil product that could be used as a recycle solvent. Realization of the second objective was a preferred solution because it would eliminate the cost of solvent recovery and solvent losses.

Of the several types of high boiling solvents tested, anthracene oil or certain oxygenated solvents (e.g., cresols and acetophenone) were preferred (Table IX). Although good results were obtained by using cresol solvents, alkylation of the cresols occurred, and it did not appear possible to recover the cresol quantitatively. In addition, stripping off the cresol left a pitchlike solid rather than an oil.

In the presence of a cresol solvent and a small amount of either sodium formate or sodium carbonate as catalyst, very little carbon monoxide is consumed during wood liquefaction at 275°C. If the amount of water is kept low, essentially no net carbon monoxide consumption occurs (Table X). Some carbon monoxide is evolved during the decomposition of the wood and is apparently about equivalent to the amount of carbon monoxide consumed. A zero net usage of carbon monoxide means that the amount in the autoclave at the end of the reaction is equal to the amount charged. The

TABLE IX

Effect of Solvent on Conversion of Sawdust to Oil[a]

	Temp. (°C)	Time (min)	Pressure (psig)		Conversion (%)	Oil (%)
			Initial	Maximum		
Anthracene oil	295	60	300	1700	97	50
Diphenyl ether	295	60	300	1800	96	45[b]
o-Cresol	295	60	300	1940	98.5	54
o-Cresol	260	60	300	1420	99.8	52
Acetophenone	275	60	300	1530	99.8	54

[a] 50 gm each of waste, water, and solvent; 5 gm Na_2CO_3.

[b] Material separated into a pitch and a thin oil.

TABLE X

Effect of Water and Solvent on Wood Liquefaction[a]

Cresol solvent (gm)	Water (ml)	Catalyst Type	Amount (gm)	Reactive pressure (psig)	Conversion (%)	Yield (%)	CO used (gm/100 gm wood)
50	200	HCO_2Na	5	1900	99.9	48	6
50	50	Na_2CO_3	2	1780	99.3	60	4
75	10	None	0	1470	99.0	58	0
75	10	HCO_2H	1	1500	99.3	58	0
75	10	HCO_2Na	1	1560	99.8	60	0

[a] 50 gm softwood sawdust, 1 hr at 275°C, 300 psig CO.

presence of a moderate carbon monoxide partial pressure, however, is beneficial because it assures adequate carbon monoxide at all times during the reaction.

The products formed at 250 and 275°C, isolated by removing the solvent by vacuum distillation, were pitchlike solids. When cresols were used, the recovered cresol fraction contained a few percent of higher alkylated phenols, indicating reaction with the cellulose degradation products.

At temperatures of 250 to 275°C the product became too thick to use after only four cycles. The data in Table XI show that temperatures of at least 300°C and operating pressures of 2700 to 3000 psig are needed to obtain a product with acceptable softening points. These data were obtained

TABLE XI

Softening Points of Softwood-Derived Bitumen[a]

Number of cycles	Temp. (°C)	Gas	Pressure (psig) Initial	Maximum	Softening point of product (°C)
5	300	CO	300	2300	32
6	300	CO	300	2400	44
7	300	CO	300	2300	104
5	300	1:1[b]	500	2250[c]	39
7	300	1:1[b]	500	2200[c]	106
7	300	CO	500	2700	46
6	325	CO	300	3000	18

[a] Rocking autoclave, 20 min at temperature, 100 ml water, 50 gm each sawdust and cresols, 5 gm each Na_2CO_3, HCO_2Na.

[b] Synthesis gas, equal parts of hydrogen and CO.

[c] The water was reduced to 33 gm and the cresol increased to 69 gm.

using carbon monoxide; higher pressures would be required if synthesis gas were used to replace the carbon monoxide because of the low activity of hydrogen at these conditions.

A low viscosity product was obtained at 325°C, but because of the high vapor pressure of water at this temperature (1750 psig), a pressure of 3000 psig was developed, starting with an initial pressure of 300 psig carbon monoxide.

IV. STOICHIOMETRY OF WOOD CONVERSION

The stoichiometry of the reaction in which cellulosic wastes are converted to oil depends largely on the severity of the processing conditions. Table I, in which cellulosic wastes are compared, shows that the carbon monoxide consumed also varies with the feedstock, which thus also has an effect on the stoichiometry. To minimize both capital and processing costs, operation at the lowest temperatures and pressures necessary to yield an acceptable product was a major objective.

Because of the expected use of wood waste in larger scale work, the stoichiometry of the conversion of wood to oil was examined at conditions that could be recommended for scale-up work. The experiment was conducted in a magnetically stirred autoclave to ensure maximum gas–liquid contact. Hardwood flour was used in this experiment because it made a slurry that stirred more easily than the softwood sawdust (both woods reacted in a similar manner). Using an initial pressure of 400 psig of 1:1 synthesis gas, a wood–water–cresol–NaCO$_3$ HCO$_2$Na ratio of 100:50:100:2.5:2.5, and 1 hr at 300°C, a 98.8% conversion and 56% oil yield were obtained. The conversions at about 1 and 20 min were 83 and 96%, respectively; the operating pressure was about 1900 psig. In the 1-hr experiment, 83.5% of the carbon, 75% of the hydrogen, and 25% of the oxygen originally in the wood were recovered in the oil. In equation form, the reaction may be expressed as

$$0.61(C_6H_{9.13}O_{4.33}) + 0.23CO + 0.08H_2 \xrightarrow[\text{catalyst}]{H_2O} H_2O + 0.64CO_2 + 0.53(C_6H_{6.93}O_{1.22}) \quad (1)$$

This equation shows only the materials that reacted and the products formed at the conditions specified and does not include water-soluble material and other losses. At higher pressures and temperatures more gas consumption can be expected. At these conditions, each 100 parts by weight of wood that reacted used a little over 7 parts of carbon monoxide and less than 0.2 part of hydrogen. Even if double these quantities of gas were required in order to improve the product characteristics, the cost of reducing gas consumed in the reaction would be relatively small.

V. ECONOMICS

Although the rapidly escalating costs of plant construction make cost prediction difficult, a set of figures developed by a chemical construction company for the Pittsburgh Energy Research Center may help to shed some light in this direction (Table XII). The figures were generated in 1974 using the best estimates of construction cost available at that time. An updated version of this material has been presented at a recent symposium [8].

In evaluating the data in Table XII, it should be remembered that the BTU content of the oil from wood wastes is not more than about 80% of the BTU content of a petroleum oil.

VI. PRODUCT CHARACTERISTICS

The oil obtained from cellulosic wastes is a dark brown viscous material. The oil obtained at temperatures of 380°C flows slowly, but the material obtained at 250°C is a soft solid that must be warmed to impart fluidity. Infrared, ultraviolet, mass spectrometry, and proton nuclear magnetic resonance (NMR) spectrometry were used to study the structure of the compounds formed. Infrared and ultraviolet spectra indicate that the oil from

TABLE XII

Summarized Economics for Conversion of Urban and Wood Wastes[a]

Urban wastes	1500 tons/day
Wood wastes	1500 tons/day
Total wastes utilized	3000 tons/day
Waste-to-oil production	2082 tons/day
Waste for synthesis gas and heat	918 tons/day
Oil production rate	3618 barrels/day
Operating costs	$9.48/ton waste

Assumed value of oil ($/bbl)	Break-even disposal charge ($/ton of wastes)
6.00	2.24
7.00	1.04
8.00	−0.17
9.00	−1.37
10.00	−2.58

[a] Total wastes contain 52.14% dry organics, 3.59% ash, and 44.00% water.

cellulose is almost entirely aliphatic with ether linkages and carbonyl and hydroxy groups present. Nuclear magnetic resonance indicated that much of the material appeared to exist in cyclic structures, and that most of the hydrogen in the product was in methyl or methylene groups. A large proportion of these latter groups were alpha or beta to a carbonyl group or to an unsaturated carbon atom. About 4% of the hydrogen was "unsaturated" hydrogen, probably olefinic rather than aromatic. Another 3% of hydrogen occurred in the form of OH groups. Inasmuch as nearly all of the original hydrogen in cellulose occurred in —OH groups, extensive changes in the original structure have occurred. There was no indication of aldehydic hydrogen in the product.

The product from wood was not examined as extensively as the product from cellulose. The wood-derived oil can be expected to be more aromatic as a consequence of the lignin present.

The BTU content of the oils and bitumens prepared from cellulose or wood ranged from 13,000 to 17,000 BTU/lb. This is less than the BTU content of petroleum oils because of the oxygen content of the wood-derived oils.

In all the work to date, the products obtained at temperatures below 300°C have been semisolid or solid materials at room temperature. At 300°C and above, the products were more fluid. Highly viscous products may be formed even at higher temperatures, if inadequate reducing gas, solvents, water, or catalyst are present.

VII. MECHANISTIC STUDIES

To gain some insight into the mechanism by which carbon monoxide converts carbohydrates to oil, the reactions of simple model compounds with carbon monoxide and with hydrogen were compared. Early work [3] showed hydrogen to be a more effective reducing agent in the absence of catalysts (other than a possible small effect of the autoclave wall) than carbon monoxide for alcohols, ketones, aldehydes, ethers, and olefins. Carbohydrates, however, were more reactive with carbon monoxide. An investigation of catalysts for the conversion of sugars to oil showed that alkaline compounds, such as sodium carbonate, were preferred catalysts, and carbon monoxide the preferred reducing agent.

The difference in behavior between carbon monoxide and hydrogen upon benzaldehyde was more striking. At 250°C with an initial carbon monoxide pressure of 1500 psig and in the presence of aqueous sodium carbonate, a 91% yield of benzyl alcohol was obtained in 1 hr. When hydrogen was used in place of carbon monoxide, the yield of benzyl alcohol dropped to 6%.

Ketones were less susceptible to reduction with carbon monoxide than aldehydes. Methyl ethyl ketone gave a 15% yield of *sec*-butanol in 1 hr at 250°C. The yield was increased to 40% by raising the autoclave temperature to 300°C.

A mechanism for the conversion of carbohydrates to oil, consistent with all the observed data, consists of the following steps:

(1)   Reaction of sodium carbonate and water with carbon monoxide to yield sodium formate:

$$Na_2CO_3 + 2CO + H_2O \rightarrow 2HCO_2Na + CO_2 \qquad (2)$$

(2)   Dehydration of vicinal hydroxy groups in a carbohydrate to an enol, followed by isomerization to a ketone:

$$-CH(OH)-CH(OH)- \rightarrow -CH=C(OH)- \rightarrow -CH_2-CO- \qquad (3)$$

(3)   Reduction of the newly formed carbonyl group to the corresponding alcohol with formate ion and water:

$$HCO_2^- + -CH_2-C(O)- \rightarrow -CH_2-CH(O^-)- + CO_2 \qquad (4)$$

$$-CH_2-CH(O^-)- + H_2O \rightarrow -CH_2-CH(OH)- + OH^- \qquad (5)$$

(4)   The hydroxyl ion, or its equivalent in terms of sodium hydroxide or carbonate, then reacts with additional carbon monoxide to regenerate the formate ion:

$$OH^- + CO \rightarrow HCO_2 \qquad (6)$$

There are, of course, many side reactions, and the final product is a complex mixture of compounds. Two types of side reactions appear to predominate, one beneficial, one undesirable.

The beneficial side reaction occurs when two or more carbonyl groups are present in one molecule. Under alkaline conditions, carbonyl groups migrate along the carbon backbone. When two carbonyl groups become vicinal, a benzylic acid type of rearrangement occurs, yielding a hydroxy acid. The hydroxy acid readily decarboxylates, and the net effect is a reduction of the remainder of the carbohydrate-derived molecule:

$$R-CO-CO-R' \xrightarrow[H_2O]{OH^-} RR'COHCO_2H \rightarrow RR'CHOH + CO_2 \qquad (7)$$

This reaction is beneficial because it converts carbohydrate-type structures to reduced compounds, eventually leading toward more paraffin-type structures having much less oxygen than the original compounds. This reaction occurs by disproportionation and does not consume carbon monoxide or hydrogen. This is the reaction believed to occur in Berl's work [6,7] using sodium hydroxide.

The other major side reaction is an aldol condensation between two carbohydrate molecules which leads to high-molecular-weight, viscous materials. This condensation reaction occurs between a carbonyl group on one molecule and two active hydrogens on another molecule with the elimination of water. In the absence of carbon monoxide or an equivalent reducing agent, this condensation becomes the major reaction and results in the formation of char. The function of the carbon monoxide is to keep the carbonyl content of the reactant system sufficiently low that liquids instead of solids are formed.

The lignin fraction of wood is also converted to an oil by processing in the presence of carbon monoxide and water. The major effect of the carbon monoxide is to reduce the carbonyl groups, via the formate [Eq. (4)], and thus eliminate much of the cross-linking and charring that normally accompany the heat treatment of lignin. Elimination of condensation and cross-linking reactions permits the stabilization of the lower molecular weight compounds formed by the breakdown of the unstable lignin molecule under the reaction conditions of the process.

VIII. CONCLUSION

Wood wastes can be converted to a bitumen or heavy oil by processing with water, sodium carbonate, and a gas rich in carbon monoxide at temperatures of 250 to 400°C and pressures of 1500 to 3500 psig. This capability is essential, for liquid fuels which can be stored and transported are the most versatile energy forms. They can be used at the site of production, or moved considerable distances from the waste-to-fuel plant.

In an effort to bring this process closer to commercialization, a small pilot plant has been built in Albany, Oregon. This facility is being tested by the U.S. Energy Research and Development Administration. It has been designed to convert 1 ton of wood chips into heavy oil daily. Success of this concept will result in another approach to the conversion of useless wastes into useful fuels.

REFERENCES

1. H. R. Appel, I. Wender, and R. D. Miller, Solubilization of low rank coal with carbon monoxide and water, *Chem. Ind.* (*London*) **47,** 1703 (1969).
2. H. R. Appell, I. Wender, and R. D. Miller, Conversion of urban refuse to oil, p. 5. U.S. Bur. Mines, Tech. Prog. Rep. 25, May 1970.
3. H. R. Appell, Y. C. Fu, S. Friedman, P. M. Yavorsky, and I. Wender, Converting organic wastes to oil: a replenishable energy source, p. 20. U.S. Bur. Mines, Rep. of Investigations 7560, 1971.

4. H. R. Appell, Y. C. Fu, E. G. Illig, F. W. Steffgen, and R. D. Miller, Conversion of cellulosic wastes to oil, p. 28. U.S. Bur. Mines, Rep. of Investigations 8013, 1975.
5. H. R. Appell, E. C. Moroni, and R. D. Miller, COSTEAM liquefaction of lignite (Preprints, Div. Fuel Chem.), *ACS Monogr.* **20**(1), 58–65 (1975).
6. E. Berl, *Naturwissenschaften* **20**, 652 (1933).
7. E. Berl, *Science* **99**, 309 (1944).
8. E. DelBel, S. Friedman, and P. M. Yavorsky, Presented at *Symp. Comparative Economics Synthetic Fuels Processing, I&EC Div., ACS, 171st National Meeting, New York, 1976.*
9. H. E. Fierz-David, *Chem. Ind. (London)* **44**, 942 (1925).
10. F. Fischer and H. Schroder, Hydrogenation of coal with carbon monoxide, *Brennst. Chem.* **2**, 257–261 (1921).
11. H. Heinemann, *Pet. Refiner* **29**(2), 111 (1950).
12. H. Heinemann, *Pet. Refiner* **33**(7), 161 (1954).
13. H. Heinemann, *Pet. Refiner* **33**(8), 135 (1954).
14. R. Weidenhagen and A. Wegner, *Ber.* **71B**, 2712 (1938).

Chapter IX

FUELS FROM WOOD WASTE

Fred Shafizadeh
DEPARTMENT OF CHEMISTRY
UNIVERSITY OF MONTANA
MISSOULA, MONTANA

I. INTRODUCTION

The heat of combustion of different types of wood, bark, and foliage is related to their chemical composition. Wood contains mainly holocellulose, whereas bark and foliage contain large amounts of phenolic and lipid extractives. The heats of combustion of these materials range from −4150 cal/gm for cellulose to −6100 cal/gm for lignin and −8400 cal/gm for the lipid extractives. The chemical composition also affects the rate of heat release and the possibility for conversion of the forest fuels to various forms of gaseous, liquid, and solid (charcoal) fuels. The carbohydrates and lipids produce mainly combustible volatiles which support the flaming combustion. Lignin, however, contributes mainly to char which burns with glowing combustion. The gasification products of cellulosic materials have generally a lower heat of combustion of about -3600 cal/gm, while the charred residue has a much higher heat of combustion of about -7100 cal/gm. Charcoal having a low sulfur content and high calorific value compares favorably with other types of solid fuel. Furthermore, it could be obtained in combination with interesting chemical by-products.

A. Wood as an Energy Source

Until the late nineteenth century, wood and wood-based fuels provided the principal energy support for industry. It is well documented that charcoal, made from wood, fueled the metal smelters as early as 3500 B.C. The island Crete rose to prominence not only because of its location but also because it had ample deposits of metalliferous ores plus forests and charcoal technology. Wood supplied heat and hot water to homes as well as industry. In Ralph Waldo Emerson's day, the wood harvested for energy in the United States was 150 million cords—over four cords per capita— equivalent to about 2 quadrillion BTU [1]. By 1880 this energy source approached 3 quads. Wood harvested specifically for fuel supplied 1.2 quads in 1950 [2].

Wood is a flexible energy source. During World War II, automobiles and buses in Finland were fitted with small reactors which converted wood into transportation fuels. Although many passengers had to climb hills when the fuel was insufficient to power loaded buses, the principle of converting wood and wood residues into clean, useful fuels was well demonstrated. Today that principle is returning as a method for increasing the energy independence of the silvicultural-based industries as they seek to decrease their consumption of purchased fuels while meeting strict environmental regulations.

With growing public awareness of the shortage of domestic energy supplies, wood residues have become significant. Forest products companies now use some 1.1 quads of energy from their wood residues [3]. These fuels have raised the pulp and paper industry from a position of 37% energy self-sufficient to a position of 42+% energy self-sufficient [4]. The plywood, veneer, and lumber industries are moving in the same direction.

B. Available Wood Residues

Wood, and wood residue, offers processors an extremely complex and diverse mixture of substances and compounds as observed in Chapter I. At first glance, the composition includes bark, sawdust, chips, ends, and other materials.

Some 25% of all wood residues produced is bark; this segment results from mills debarking logs before sawing or chipping them to increase roundwood recovery [5]. Bark presents difficult problems for utilization. It consumes more chemicals than wood in the production of pulp, and is thus expensive to use for that purpose. It has limited application as a soil conditioner, basically in local markets.

Sawdust is the second most problematical waste for wood-based industries. This material does not have the same long fibers offered by chips. Thus it is less useful in the manufacture of particle board and flakeboard, although some sawdust can be consumed by those industries.

Chips, shavings, and other waste or residue materials have available markets in the paper industry and the particle board manufacturing community. They are, at this point, by-products of sawmills and planing mills. Classification of these materials as residues or wastes is no longer economically accurate.

Considerations of using the various wood residues involve not only the direct combustion of natural fuels and cellulosic wastes, but also their conversion to various types of gaseous, liquid, and solid fuel, with or without the production of chemical by-products, as well as the conservation of forest resources and prevention of destructive forest fires. The following discussions on chemical composition, pyrolysis, combustion, and the heat content of the natural fuels [6–10] provide a technical basis for the related developments.

II. CHEMICAL COMPOSITION

The wood residues and related natural fuels contain lignocellulosic materials which form the plant cell walls, absorbed and condensed moisture, various extractives, and some mineral compounds. These components could vary for different parts and species of the plant. However, on a dry basis, the wood substance contains about 7% of extractives and minor amounts of minerals or ash. Extracted hardwoods contain about 43% cellulose, 35% hemicelluloses, and 22% lignin, whereas softwoods contain about 43% cellulose, 28% hemicelluloses, and 29% lignin.

Acetyl-4-o-methylglucuronoxylan ("xylan") forms the main hemicellulose of the hardwoods and glucomannans ("mannan") form the principal hemicelluloses of the softwoods. Furthermore, the softwood lignin contains guaiacyl propane units (having one methoxyl group) and hardwood lignin, in addition to this, contains syringyl propane units (with two methoxyl groups). The higher content of acetyl and methoxyl groups in hardwoods explains why this material has been used in destructive distillation processes to obtain acetic acid and methanol.

TABLE I

Analysis of Douglas Fir Wood, Bark, and Needles

Fraction	Wood (%)	Bark (%)	Needles (%)
Ether soluble	1.2	34.4	11.2
Benzene–alcohol soluble	4.4	—	19.1
Hot water soluble	5.6	29.2	7.2^a
Ash	0.2	0.9	5.6
Lignin and phenolics	27.2	26.1	36.4
Holocellulose	61.2	9.4^a	20.5

^{a} By difference.

In comparison to the wood substance, bark contains much more extractives and lignin or phenolic compounds, which have a higher heat of combustion, and a lower amount of holocellulose (combination of cellulose and hemicellulose), which has a lower heat of combustion. The needles and leaves also have a higher percentage of extractives as shown in Table I.

III. PYROLYSIS AND COMBUSTION

The lower molecular weight ether extractives, particularly the terpenoid hydrocarbons (turpentine components), have a high calorific value and readily evaporate and burn in the gas phase with flaming combustion.

Lignocellulosic materials, however, are not directly combustible, but under the influence of a sufficiently strong source of energy or pilot ignition, decompose to form volatile pyrolysis products, which burn in the gas phase with flaming combustion. The charred residue burns at a relatively slower rate by surface oxidation or glowing combustion.

In these transformations, shown in Fig. 1, the cellulosic component is mainly converted to the combustible and noncombustible volatiles, including water and carbon dioxide, whereas the lignin component contributes mainly to the char fraction. The initial depolymerization of cellulose produces an intermediate tar fraction containing levoglucosan.

The rate of gasification and the amount of residue left at different temperatures could be determined by thermal analysis as shown for cellulose (Fig. 2). Thermal analysis of cottonwood and its components shown in Figs. 3 and 4 indicates that pyrolysis of the cellulose and hemicellulose components gives mainly volatile products, whereas lignin gives mainly char. Furthermore, the pyrolysis of wood reflects the thermal properties of its components.

Thermal degradation of cellulose and hemicelluloses to flammable volatile products and chars involves a series of highly complex reactions and a variety of products, which have been extensively investigated in the author's laboratory [8,10]. These reactions, which take place both concurrently and consecutively, may be classified in the following categories:

(1) Depolymerization of the polysaccharides by transglycosylation at about 300°C to provide a mixture of levoglucosan, other monosaccharide derivatives, and a variety of randomly linked oligosaccharides, as shown in Fig. 5 and Table II for cellulose [10]. This mixture is generally referred to as the tar fraction.

(2) These reactions are accompanied by dehydration of sugar units in cellulose, which give unsaturated compounds, including 3-deoxyglucosenone, levoglucosenone, furfural, and a variety of furan derivatives, which are found partly in the tar fraction and partly among the volatiles.

(3) At somewhat higher temperatures, fission of sugar units provides a variety of carbonyl compounds, such as acetaldehyde, glyoxal, and acrolein, which readily evaporate.

(4) Condensation of the unsaturated products and cleavage of the side chains through a free radical mechanism leave a highly reactive carbonaceous residue containing trapped free radicals.

Heating of the cellulosic materials at or above 500°C provides a mixture of all these products, as shown in Tables III and IV for cellulose and xylan.

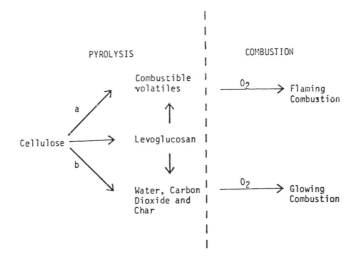

Fig. 1. Competing reactions in the pyrolysis and combustion of cellulose.

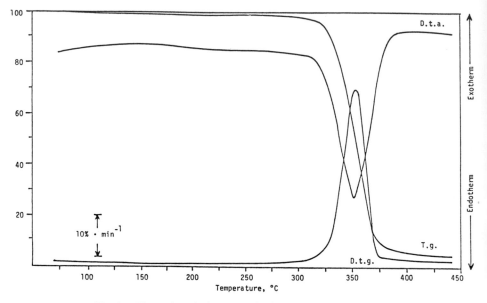

Fig. 2. Thermal analysis curves of microcrystalline cellulose.

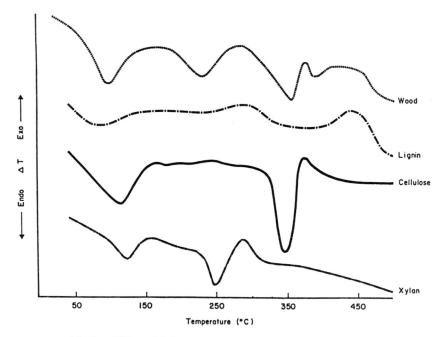

Fig. 3. Differential thermal analysis of wood and its components.

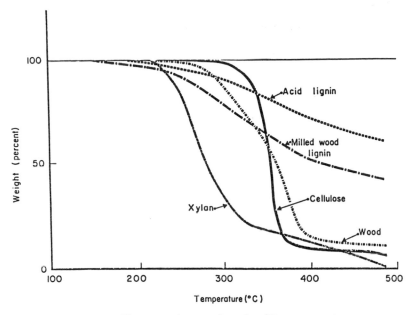

Fig. 4. Thermogravimetry of wood and its components.

Addition of an acidic catalyst or slow heating promotes the dehydration and charring reactions. Therefore, higher temperature, faster heating rate, and smaller particle size promote the gasification process and rapid flaming combustion, whereas lower temperatures, larger particle size, and the presence of moisture and inorganics increase the amount of char and favor smoldering combustion or no combustion at all.

In this connection, it should also be noted that the hydrocarbons and lipids classified as the ether extractives not only burn more rapidly and

TABLE II

Analysis of the Pyrolysis Products of Cellulose at 300°C under Nitrogen

Condition	Atmospheric pressure	1.5 mm Hg	1.5 mm Hg, 5% SbCl₃
Char (%)	34.2	17.8	25.8
Tar (%)	19.1	55.8	32.5
levoglucosan	3.57	28.1	6.68
1,6-anhydro-β-D-glucofuranose	0.38	5.6	0.91
D-glucose	Trace	Trace	2.68
hydrolyzable materials	6.08	20.9	11.8

TABLE III

Pyrolysis Products of Cellulose and Treated Cellulose at 550°C

Product	Neat	+5% H_3PO_4	+5% $(NH_4)_2HPO_4$	+5% $ZnCl_2$
Acetaldehyde	1.5[a]	0.9	0.4	1.0
Furan	0.7	0.7	0.5	3.2
Propenal	0.8	0.4	0.2	Trace
Methanol	1.1	0.7	0.9	0.5
2-Methylfuran	Trace	0.5	0.5	2.1
2,3-Butanedione	2.0	2.0	1.6	1.2
1-Hydroxy-2-propanone Glyoxal	2.8	0.2	Trace	0.4
Acetic acid	1.0	1.0	0.9	0.8
2-Furaldehyde	1.3	1.3	1.3	2.1
5-Methyl-2-furaldehyde	0.5	1.1	1.0	0.3
Carbon dioxide	6	5	6	3
Water	11	21	26	23
Char	5	24	35	31
Tar	66	16	7	31

[a] Percentage, yield based on the weight of the sample.

TABLE IV

Pyrolysis Products of Xylan and Treated Xylan at 500°C

Product	Xylan		o-Acetylxylan	
	Neat	+10% $ZnCl_2$	Neat	+10% $ZnCl_2$
Acetaldehyde	2.4[a]	0.1	1.0	1.9
Furan	Trace	2.0	2.2	3.5
Acetone Propionaldehyde	0.3	Trace	1.4	Trace
Methanol	1.3	1.0	1.0	1.0
2,3-Butanedione	Trace	Trace	Trace	Trace
1-Hydroxy-2-propanone	0.4	Trace	0.5	Trace
3-Hydroxy-2-butanone	0.6	Trace	0.6	Trace
Acetic acid	1.5	Trace	10.3	9.3
2-Furaldehyde	4.5	10.4	2.2	5.0
Carbon dioxide	8	7	8	6
Water	7	21	14	15
Char	10	26	10	23
Balance (tar)	64	32	49	35

[a] Percentage, yield based on the weight of the sample.

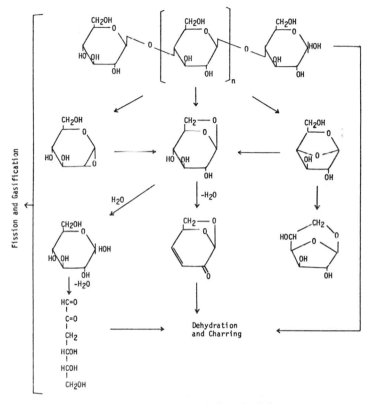

Fig. 5. Thermal degradation of cellulose.

generate more heat than the wood substance, but they also promote the rapid and more complete combustion of the lignocellulosic material by increasing the temperature and the rate of gasification.

IV.   HEAT CONTENT

The heat contents of different types of forest fuels and their components are given in Table V [11]. This table also shows the amount of char left at 400°C on heating at the rate of 200°C/min and the heat of combustion of the gaseous and carbonaceous products. Table VI shows the distribution of the heat of combustion between the char and the volatile products. These data show the relationship between the heat of combustion and the chemical composition of the fuel.

TABLE V

The Heat of Combustion of Natural Fuels and Their Pyrolysis Products as Char and Combustible Volatiles

Fuel			Char		Combustible volatiles	
Source	Type	$\Delta H^{25°}_{comb}$ (cal/gm)	Yield (%)[a]	$\Delta H^{25°}_{comb}$ (cal/gm)	Yield (%)[a]	$\Delta H^{25°}_{comb}$ (cal/gm)
Cellulose	Filter paper	−4143	14.9	−7052	85.1	−3634
Douglas fir lignin	Klason	−6371	59.0	−7416	41.0	−4867
Poplar wood						
Populus spp.	Excelsior	−4618	21.7	−7124	78.3	−3923
Larch wood						
Larix occidentalis	Heart wood	−4650	26.7	−7169	73.3	−3732
Decomposed Douglas fir						
Pseudotsuga menzeisii	Punky wood	−5120	41.8	−7044	58.2	−3738
Ponderosa pine						
Pinus ponderosa	Needles	−5145	37.0	−6588	63.0	−4298
Aspen						
Populus tremuloides	Foliage	−5034	37.8	−6344	62.2	−4238
Douglas fir bark						
Pseudotsuga menzeisii	Outer (dead)	−5122	52.8	−5798	47.2	−4366
Douglas fir bark						
Pseudotsuga menzeisii	Whole	−5708	47.1	−6406	52.9	−5087

[a] Heating rate 200°C/min to 400°C and held for 10 min.

TABLE VI

Distribution of the Heat of Combustion of Forest Fuels

Fuel		Char	Gas	Total
Source	Type	(cal/gm fuel)	(cal/gm fuel)	(cal/gm)
Cellulose	Filter paper	−1050	−3093	−4143
Douglas fir lignin	Klason	−4375	−1995	−6370
Poplar wood				
Populus spp.	Excelsior	−1546	−3072	−4618
Larch wood				
Larix occidentalis	Heart wood	−1914	−2736	−4650
Decomposed Douglas fir				
Pseudotsuga menzeisii	Punky wood	−2944	−2176	−5120
Ponderosa pine				
Pinus ponderosa	Needles	−2438	−2708	−5146
Aspen				
Populus tremuloides	Foliage	−2398	−2636	−5034
Douglas fir bark				
Pseudotsuga menzeisii	Outer (dead)	−3061	−2061	−5122
Douglas fir bark				
Pseudotsuga menzeisii	Whole	−3017	−2691	−5708

Pure cellulose, which is composed entirely of sugar units with the elemental analysis of $C_6(H_2O)_5$, has a relatively low heat content ($\Delta H^{25°}$ −4143 cal/gm), because of the high level of oxidation. However, 75% of its heat content is released to the volatiles because of the transglycosylation and thermal cleavage of the polysaccharide, discussed earlier.

Lignin, which is composed of coniferyl units (and some related syringyl units in hardwoods), with the elemental analysis of $C_{10}H_{11}O_2$, has a lower degree of oxidation and a considerably higher heat of combustion ($\Delta H^{25°}$ −6371 cal/gm). On pyrolysis, it forms mainly char because it is not readily cleaved to low-molecular-weight fragments.

The poplar and larch samples, which contain about 25% lignin, show intermediate heat of combustion and gasification characteristics.

The effect of lignin content is more pronounced with punky wood samples in which the cell wall polysaccharides are partially removed by biological degradation resulting in 52% lignin content. This sample has a still higher heat of combustion but lower heat release ratio of 41%. This is why the punky wood supports smoldering or latent combustion, which holds the fire over long periods of time, rather than the rapidly developing and consuming flaming combustion.

The ether extractives (terpenoid hydrocarbons and lipids) have a still lower oxygen and higher heat content ($\Delta H^{25°}$ − 7700–8500 cal/gm) and affect the heat of combustion of ponderosa pine and aspen foliage which have a high extractive content. This effect is complicated by the presence of some ash and various amounts of lignin, in addition to the ether extractives. However, comparison with the corresponding extracted samples showed that the absence of ether extractives not only lowers the total heat of combustion of the fuel but also lowers the ratio of heat release from 50 to 42% for aspen foliage. In other words, the extractives provide a source of readily vaporized and high energy combustible materials.

Figure 6 shows the heat of combustion of the volatiles produced by gasification of Douglas fir needles at different temperatures before and after removal of the extractives. This figure and similar data obtained for various forest fuels [9,11] dramatically demonstrate the contribution of the extractives to the combustibility of forest fuels.

These considerations indicate a definite correlation between the thermal properties of natural fuels and their chemical composition both at the molecular and atomic levels.

At the latter level, the heat content is clearly related to the oxidation state of the natural fuels in which carbon atoms generally dominate and overshadow small variations of hydrogen content. This situation leads to a rather unexpected, but highly interesting and useful correlation. When the heats of combustion of various substrates, including the fuels, chars, and volatiles, are plotted as a function of their respective carbon content (Fig. 7), the least squares line through the individual points fits the equation

$$\Delta H^{25°} \text{ combustion (cal/gm)} = 94.19 \, (\%C) + 55.01$$

This correlation is due to an averaging and mutual cancellation of different effects, particularly the effect of hydrogen and oxygen contents. Hydrogen atoms should increase the heat content above the average value and oxygen atoms should reduce it below the average value. For instance, in the extreme case of CO_2, the heat of combustion is nil, and the carbon content is 27.3%. However, within the range of chars, fuels, and volatiles that have been considered, the chars have lower hydrogen content, but also lower oxygen content; the reverse is true for the volatiles and the original fuels which are in between. In all these substrates, the two opposing effects cancel each other.

V. EVALUATION

For a realistic evaluation of forest fuel, the heats of combustion obtained from calorimetric determination at 25°C should be corrected to the ignition temperature by considering the heat of preignition required for heating the

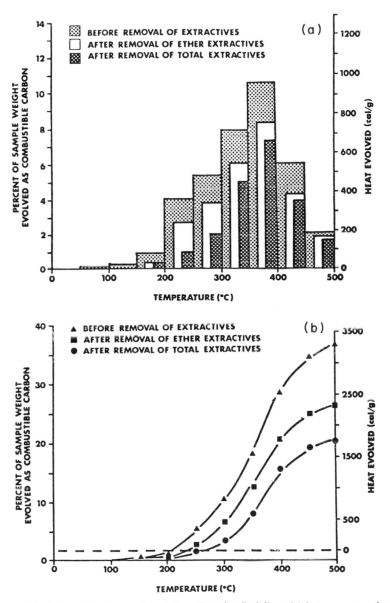

Fig. 6. Evolution of carbon and heat from Douglas fir foliage (a) in temperature intervals and (b) cumulative, based on dry weight of the unextracted sample.

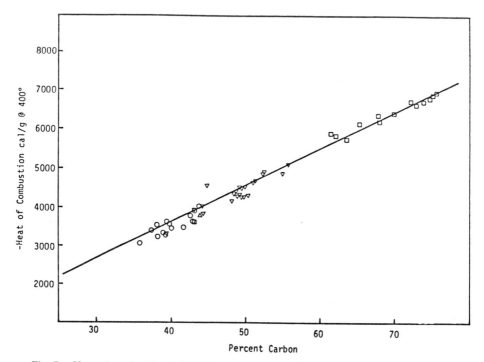

Fig. 7. Heat of combustion at 400°C versus percent carbon: ∇, fuels; $\square$, char; $\bigcirc$, volatiles.

forest fuels to the ignition temperature. The heat of preignition includes the energy required for vaporization of the moisture content (about 582 cal/gm), the heat of pyrolysis, and the heat capacities required for raising the temperature of the various products (200–250 cal/gm). Also, since the flaming combustion is fueled by production of the flammable volatiles, in the evaluation of flammability and flame spread, allowance should be made for a portion of fuel which is converted to char. As discussed previously, the char fraction burns by surface oxidation or glowing ignition that mainly provides radiation rather than convection energy.

Generally, the balance between the heat of combustion of the volatile products and the heat of preignition provides a good indication of the flammability. These values could be obtained as a function of temperature by thermal evolution analysis (TEA), which gives the heat of combustion of the volatiles, and differential scanning calorimetry (DSC), which gives the heat of preignition [13]. The results could be used in the following equation for modeling of the fire behavior [14].

$$I_{\rm R} = -h \, \frac{dw}{dt}$$

TABLE VII

Yield of Pyrolysis Products from 100 kg of Softwood

Product	Yield (kg)	Product	Yield (kg)
Charcoal	32	Turpentine oil	0.6
A-Tar	7	Light oil	0.4
B-Tar	3	Methanol	1.0
Acetic acid	1.7	Uncondensable	22.0
Acetone	0.8	gases	

where I_R is the reaction intensity, h the heat of combustion, and dw/dt the rate of mass loss.

Wet fuels not only have a lower heat of combustion per unit weight at 25°C but also a higher heat of preignition and produce more char which reduces the intensity of the flaming combustion. The combined effects could result in self-extinguishment or blackout. Consequently, in designing wood-burning furnaces, it is advantageous to use the heat content of the flue gases for predrying the fuel.

On a weight per weight basis, the average heat value of natural fuels, 4500 cal/gm, is not too far below the average heat values for different types of coal, which range from 7915 cal/gm for bituminous grade A to 3640 cal/gm for lignite. The comparison becomes more favorable by pyrolysis or destructive distillation of natural fuels, which could provide charcoal and various chemical by-products. The low sulfur content, relatively high heat of combustion, and an inherently low production cost provide the main incentives for conversion of the cellulosic wastes and natural fuel to charcoal. Destructive distillation of resinous softwood in addition to charcoal could provide turpentine and pine tar (see Table VII), whereas hardwoods and agricultural residues could provide furfural, methanol, and acetic acid as by-products (see Table VIII).

TABLE VIII

Destructive Distillation of Hardwoods

Products	Yield (% of dry wood)
Charcoal (17.5% volatiles)	36
Tars and oils	12
Noncondensable gases	20
Pyroligneous acid, containing	32
Water and minor amount of miscellaneous compounds	24
Acetic acid (including formic and propionic acids)	6
Methanol and acetone	2

The old destructive distillation method could be modernized to provide for the increasing need for various types of fuel. The old, established producer gas process, based on the following reactions, could also be modernized with the application of new engineering technology:

$$natural\ fuel \rightarrow C + H_2O$$
$$C + O_2 \quad \rightarrow CO_2$$
$$CO_2 + C \quad \rightarrow 2CO$$
$$H_2O + C \quad \rightarrow CO + H_2 \Big\} combustible\ gas$$

VI. COMMERCIALIZATION

A review and detailed discussion of these processes are beyond the scope of this chapter. At the same time, however, certain systems merit brief mention.

A. Fixed-Bed Systems

Based on research performed at the Engineering Experiment Station of the Georgia Institute of Technology, a relatively low temperature pyrolysis system has emerged. Tech-Air Corporation, a wholly owned subsidiary of the American Can Company, has taken the Georgia Tech process to the brink of commercialization. Their installation is a 50-ton/day unit at Cordele, Georgia. It pyrolyzes sawdust, bark, and partially decomposed bark.

The Tech-Air system consists of a vertical reactor operated at 1100°F to favor production of solid fuels. Lesser amounts of liquid and gaseous fuels are also produced. A portion of the gaseous fuel produced is used to dry the incoming waste to 4% moisture. The char produced has a heat content of 11,000 to 13,500 BTU/lb, the oil has a heat content of 10,000 to 13,000 BTU/lb, and the gas offers 200 BTU/ft³ [15]. This system, depicted in Fig. 8, has been field tested for two years and has operated successfully since 1975.

Georgia Tech has also developed a mobile unit for pyrolysis of agricultural and silvicultural residues. It operates on the same basic design as the Tech-Air system.

Other fixed-bed systems which can be applied to the production of superior fuels from wood waste includes the Purox system of Union Carbide Corporation, developed initially for municipal waste. It produces a gas containing 350 BTU/ft³.

B. Fluidized-Bed Gasification

At Morgantown, West Virginia, university researchers employed a fluidized-bed system for wood waste gasification. The reactor was designed

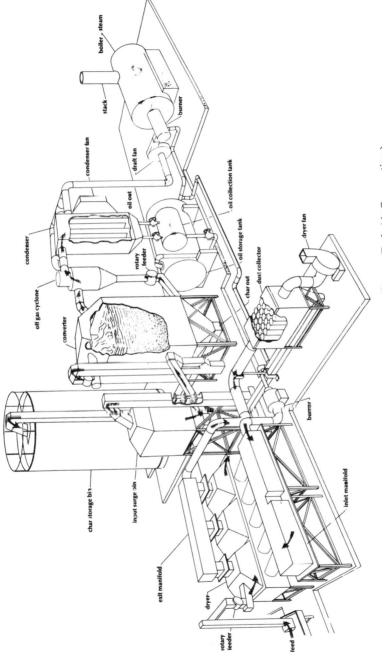

Fig. 8. How the Tech-Air system works. (Source: Tech-Air Corporation.)

to handle a wide variety of carbonaceous feedstocks [16]. The system employed consisted of a 15-in. (i.d.) fluidized-bed chamber with an 8-ft height. Beneath the reactor was an L-shaped hot bottom chamber where gas was burned. Combustion gases passed through a high temperature stainless steel grid plate with 584 holes, each 0.096 in diameter, located on 0.25-in. centers. Solid feeding was accomplished by a screw feeder 5 in. above the grid plate. Above the reactor was a 22-in. (i.d.) gas expansion zone followed by analysis equipment and gas cleaning apparatus.

The basic element of the system, the fluidized bed, was filled with sand to a height of 30 in. When the system operated, this bed expanded to a height of between 42 and 48 in. It exposed new reactive surfaces of the sawdust by abrasion and grinding, provided very rapid heat transfer to the wood waste particles, and acted as a heat sink to minimize temperature fluctuations.

The fluidized-bed experiments demonstrated that over 80% of the energy contained in the sawdust could be converted into an excellent fuel gas containing up to 12.4% CH_4 and up to 4.7% C_2H_x compounds. This gas provides the user with 286 to 412 BTU/scf. Thus it can be used in natural-gas-burning systems with little modification.

A third approach, catalytic gasification, may emerge from the research designed to produce oil from wood waste. This research is described in Chapter VIII. Subsequent processes which can be employed to re-form the gaseous products into methanol are described in Chapter XI.

VII. CONCLUSION

Wood and wood waste, the original fuel which helped man achieve civilization, are again gaining consideration as an energy source. The chemical composition of this material, plus specific process configurations, determines the specific materials to be produced.

Although wood and wood waste may be combusted directly, they can also be converted into superior products: char and charcoal, gas, and oil.

REFERENCES

1. K. Kern, Wood power: heating and cooking with wood, *in* "Producing Your Own Power" (C. H. Stoner, ed.). Rodale Press, Emmaus, Pennsylvania, 1974.
2. S. H. Schurr and B. C. Netschert, "Energy in the American Economy 1850–1955." The Johns Hopkins Press, Baltimore, Maryland, 1960.
3. K. V. Sarkanen, Renewable resources for the production of fuels and chemicals, *Science* **191** (4228), 773 (1976).
4. Problems and legislative opportunities in the basic materials industries. National Materials Advisory Board, National Academy of Sciences, 1975.

5. E. P. Cliff, Timber: the renewable resource. National Commission on Materials Policy, August 1973.
6. F. Shafizadeh, *Adv. Carbohydr. Chem.* **23,** 419 (1968).
7. F. Shafizadeh, *Pure Appl. Chem.,* **35,** 195 (1973).
8. F. Shafizadeh, *Appl. Polym. Symp.* **28,** 153 (1975).
9. F. Shafizadeh, *J. Polym. Sci., Part C.* **36,** 21 (1971).
10. F. Shafizadeh and Y. L. Fu, *Carbohydr. Res.* **29,** 113 (1973).
11. R. A. Susott, W. F. DeGroot, and F. Shafizadeh, *J. Fire Flammability* **6,** 311 (1975).
12. F. Shafizadeh, P. P. S. Chin, and W. F. DeGroot, *J. Forensic Sci.,* in press.
13. F. Shafizadeh, P. P. S. Chin, and W. F. DeGroot, Fire retardant chemistry, *J. Fire Flammability,* **2,** 195 (1975).
14. R. C. Rothermel, USDA Forest Serv., Res. Paper INT-115, 1972.
15. J. A. Knight, Pyrolysis of pine sawdust, *in* "Thermal Uses and Properties of Carbohydrates and Lignins" (F. Shafizadeh, K. V. Sarkanen, and D. A. Tillman, eds.). Academic Press, New York, 1976.
16. C. Y. Wen *et al.,* Production of low BTU gas involving coal pyrolysis and gasification, *in* "Coal Gasification" (Lester G. Massey ed.). Advances in Chemistry Series 131, Amer. Chem. Soc., Washington, D.C., 1974.

Chapter X

PYROLYTIC GASIFICATION OF KRAFT BLACK LIQUORS

K. T. Liu, E. P. Stambaugh, H. Nack, and J. H. Oxley
BATTELLE-COLUMBUS LABORATORIES
COLUMBUS, OHIO

I. INTRODUCTION

For every ton of pulp produced in the Kraft process, approximately 1.6 tons of black liquor solids are also produced. These solids consist of 60% organic material and 40% inorganics. In this country, of the 36 million tons of chemical pulp produced, 32 million tons are made by the Kraft process. It has become the predominant method for liberating paper fibers from pulp wood. Thus, today some 51.2 million tons of black liquor solids are generated annually, of which 31 million tons are organic matter [1].

Currently, black liquor supplies the pulp and paper industry with some 0.675 quadrillion BTU/year of energy (Table I). The spent liquor is combusted in special furnace systems where energy and inorganic chemicals required by the pulping process are recovered. All energy is recovered as steam which is used internally. The pulp and paper industry produces

TABLE I

Use of Residues as Fuel in the Pulp and Paper Industry[a]

Residue used	BTU provided (in 1×10^{12})	
	1975	1972
Hogged fuel	56.1	41.1
Bark	82.8	109.1
Spent liquor	675.8	759.1
Total	814.7	909.3

[a] Source: American Paper Institute.

between 42 and 45% of its energy needs from burning residues [2,3]. Some 35% of the total energy requirement in this industry comes from black liquor combustion [4]. It is the single most important source of energy in the pulp and paper industry [4]. Table I depicts that situation. Thus, black liquor is also the most utilized energy-producing residue in the U.S. economy.

The potential for using black liquor has not been reached. Between 1972 and 1975 the volume used as an energy source declined from approximately 60 million tons to 54 million tons, or 10%. Numerous reasons underlie this fact, among them being the form of energy resulting from black liquor. Steam is not a storable or transportable fuel. For a waste to be more completely used as a fuel it must be converted into a solid, liquid, or gaseous form.

Research performed at the Pittsburgh Energy Research Center demonstrated that black liquor could be combined with bark and pelletized for pyrolysis [5]. This offers one approach to the problem.

Goheen et al. [1] suggest that a portion of the liquor could be pyrolyzed to produce guaiacol. The carbon residue from this process would be dispersed throughout the unprocessed liquor. Energy recovery could proceed with less than a 20% loss in fuel value.

Catalytic gasification offers another, highly useful approach. The end result would be a fuel gas which could be combusted directly or converted into methanol as discussed in Chapter XI. At the same time, steam could be supplied to the plant.

II. THE CONCEPT

In the United States, much of the spent liquor is concentrated and burned to provide part of the energy for the plant. One potential method of utilizing these liquors is to produce gaseous products for fuels, chemicals, and other applications. This suggestion is based on the following facts:

(1)   The carbonaceous material of black liquor is well dispersed.

(2)   The black liquor contains a considerable amount of alkali metals, which are known to have excellent catalytic effect on gasification.

(3)   The concentrated black liquor contains about 40 to 60% water, which can be converted to steam for the carbon–steam gasification reaction.

Because of the presence of carbonaceous material and water in the black liquor, it is suspected that the following water–carbon reactions would predominate:

$$C + H_2O \rightarrow CO + H_2 \tag{1}$$

$$CO + H_2O \rightarrow CO_2 + H_2 \tag{2}$$

Normally, the hydrogen concentration in the product gas should not exceed a certain limit imposed by thermodynamic equilibrium. For example, for a typical sodium-base spent liquor, the hydrogen concentration in the pyrolytic gasification products at 1000 K and ambient pressure will not exceed 60% in volume [reaction (1)]. However, in the presence of a CO_2-removal reagent, such as NaOH or CaO, the equilibrium can be shifted to drive reaction (2) toward completion. This would maximize the hydrogen yield, and at the same time reduce CO and CO_2 concentrations in the product gas.

The feasibility of utilizing the spent liquor via pyrolytic gasification at nearly atmospheric pressures has been demonstrated by Prahacs and Gravel [6] and Prahacs [7]. They concluded that the sodium-base liquor gave highest yields of hydrogen and carbon monoxide. However, the hydrogen in the product gas was generally in the range 50 to 60% by volume.

The objective of this chapter is to demonstrate experimentally the technical feasibility of producing hydrogen-rich gas by pyrolytic gasification of black liquor. Sodium hydroxide was selected as a CO_2-removal reagent for convenience of handling in the microreactor used in this study.

III.   RESEARCH EXPERIMENTS AND RESULTS

To examine the feasibility of this approach, numerous experiments were conducted. These demonstrated the utility of the concept.

A.   Experiments

The gasification experiments were conducted in a batch reactor system as shown in Fig. 1. The reactor was made of a 12-in. long ¾-in. Inconel pipe

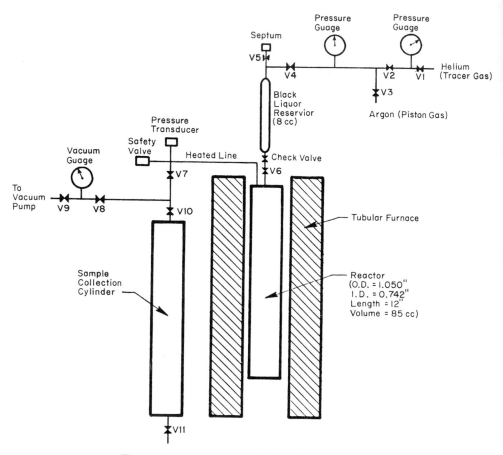

Fig. 1.   Experimental setup for black liquor gasification.

(i.d. 0.742 in., o.d. 1.050 in.). Helium was used as an inert tracer to provide a material balance from which gasification yield was estimated.

In a typical experiment, the reactor was brought to the selected reaction temperature, evacuated, and then pressurized to 20 psig with helium. With valves V6 and V7 closed, a measured amount of water (0.5 ml, typically) was loaded into the sample reservoir through V5 by means of a hypodermic syringe and V5 was closed. The free space above the water sample in the reservoir was pressurized to 300 psig with argon (to serve as piston gas) and isolated by closing V4 before the water was forced into the heated reactor by opening valve V6 and immediately closing it. This provided a steam environment for gasification. With the same procedure, a measured amount of black liquor (2.0 ml) was then injected into the steam-filled reactor, except that 600 psig of helium pressure was used as a piston gas. Vaporization of the solution took place almost instantaneously. After the predetermined reaction period, the products were collected by opening valves V7 and V10 to transfer the sample to the collection cylinder. The collected gaseous reaction products were analyzed by gas chromatography and mass spectroscopy.

All experimental tests discussed in this chapter were conducted at 800°C. The reaction time was chosen to be 2 min.

The sodium-base black liquor used in this study contained 15.6% by weight organic carbon. To investigate the addition of CO_2-removal reagent, various amounts of sodium hydroxide were added to the liquor at Na mole ratios of 0.00, 0.38, 0.77, and 1.15, where Na is moles of added sodium and C is moles of organic carbon in the liquor.

B. Results

The results from these experiments are shown in Table II and Figs. 2 and 3. The degree of gasification was estimated by assuming that hydrogen and

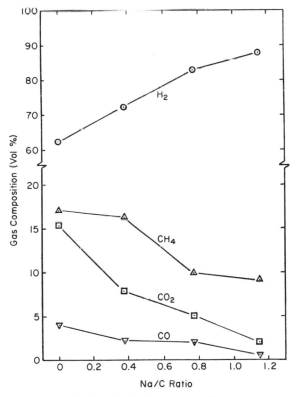

Fig. 2. Product gas composition.

carbon dioxide are produced according to the carbon steam reaction

$$C + 2H_2O \rightarrow CO_2 + 2H_2$$

Therefore, for every 2 moles of hydrogen produced, 1 mole of organic carbon in the black liquor should be gasified. The carbon dioxide thus formed would react with NaOH to form sodium carbonate, up to the limit imposed by the quantity of NaOH added and/or already present in the black liquor. The following observations were made: The main reaction products were found to be hydrogen, methane, carbon dioxide, and carbon monoxide; very small amounts of C_2–C_6 hydrocarbons (rarely exceeding

TABLE II

Results of Pyrolytic Gasification of Black Liquor

Run No.	PG-01	PG-02	PG-03	PG-04
Temperature (°C)	800	800	800	800
Na/C mole ratio in feed[a]	0.00	0.38	0.77	1.15
Product yield (gm-mole/1000 ml of as-received liquor)				
H_2	6.85	11.96	16.77	18.35
CO	0.45	0.36	0.40	0.12
CH_4	1.88	2.70	2.01	1.90
CO_2	1.69	1.30	1.00	0.42
C_2H_4	0.008	0.012	0.007	nil
C_2H_6	0.039	0.087	0.012	0.051
C_3H_6	0.002	0.007	0.003	0.006
C_3H_8	nil	0.005	nil	nil
C_6H_6	0.005	0.12	0.077	0.08
Product composition (%)				
H_2	62.4	72.3	82.7	87.7
CO	4.1	2.2	2.0	0.57
CH_4	17.1	16.3	9.9	9.1
CO_2	15.4	7.9	5.0	2.0
C_2H_4	0.07	0.07	0.03	nil
C_2H_6	0.35	0.53	0.08	0.24
C_3H_6	0.01	0.04	0.02	0.03
C_3H_8	nil	0.03	nil	nil
C_6H_6	0.6	0.72	0.38	0.38
	100.33	100.09	100.11	100.02
Percent gasification (based on carbon input)	40.0	64.0	72.5	75.6

[a] Na content refers to the amount of NaOH added to the liquor. C content is 15.6 wt %.

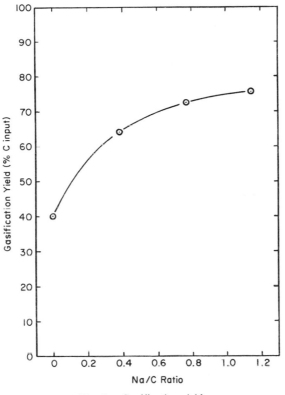

Fig. 3. Gasification yield.

1.3% total) were observed; the concentration of hydrogen in the gaseous products was very high in all experiments, ranging from 62 to 88 vol %.

There are two striking effects of the added caustic soda on the gasification of black liquor. The first is the product gas distribution, and the second is the total gasification yield. As shown in Fig. 2, the hydrogen content of the gas increased with increasing Na/C ratio; about 88% of hydrogen was obtained with a Na/C ratio of 1:15 as compared to 62% when no free NaOH (Na/C = 0.00) was added to the black liquor. Also, the CO and CH_4 concentrations, in general, decreased progressively with increasing Na/C ratio. The total gasification yield as a function of the Na/C ratio is shown in Fig. 3. Significant increases in total gasification yield with increasing Na/C ratio were observed. About 76% of conversion can be obtained at a Na/C of 1:15, as compared to 40%, when no free NaOH is added to the black liquor. This suggests that the added alkali metal greatly enhances the gasification reaction.

IV. CONCLUSIONS AND DISCUSSION

In conclusion, the production of hydrogen in concentrations greater than 85% in the gas from pyrolytic gasification of black liquor is technically feasible. The addition of caustic soda to the black liquor not only enhances the gasification reaction but also increases the hydrogen concentration in the product gas. Other reagents, such as calcined limestone or dolomite, may be substituted for the caustic soda.

A potential application of this black liquor gasification process is to produce hydrogen for ammonia synthesis. Currently, the production of ammonia consumes a significant share of the natural gas in the United States. Natural gas is in short supply, and proven reserves are declining. Thus, alternative raw materials are being sought. Coal gasification is considered to be a potential method. However, preliminary results as discussed in this chapter have indicated that black liquor can be processed to produce a gas stream with relatively higher concentration of hydrogen than that obtained from direct gasification of coal. With carbonaceous materials well dispersed, black liquor appears to represent a better gasification feedstock than coal for the production of hydrogen.

If black liquor is pyrolyzed to produce an exportable product, the pulp and paper industry will have to replace it as an energy source—either partially or totally. This can be accomplished either by purchasing energy or, more probably, by increasing the combustion of bark and hogged fuel. Combustion offers one approach. The production of fuel gas, discussed in Chapter IX, provides another method for the pulp and paper industry to use forest residues in place of the gasified black liquor.

Thus, a waste material, which presents a disposal problem and a potential pollution hazard, can be converted to products for useful fuels, chemical feedstocks, and/or applications.

REFERENCES

1. D. W. Goheen *et al.*, Indirect pyrolysis of Kraft black liquor, *Amer. Chem. Soc. Meeting, September 1-2, 1976,* p. 1.
2. Problems and legislative opportunities in the basic materials industries, p. 54. National Materials Advisory Board, National Academy of Sciences, 1975.
3. The feasibility of utilizing forest residues for energy and chemicals. National Science Foundation, March, 1976.
4. J. M. Duke, Patterns of fuel and energy consumption in the pulp and paper industry. American Paper Institute, March, 1974.
5. M. D. Schlessinger *et al., IITRI Proceedings,* 1972.

6. S. Prahacs and J. J. O. Gravel, Gasification of organic content of sodium-base spent pulping liquors in an atomized suspension technique reactor, *Ind. Eng. Chem. Proc. Des. Dev.* **6** (2), 180 (167).

7. S. S. Prahacs, Pyrolytic gasification of Na-, Ca-, and Mg-base spent pulping liquors in an AST reactor *in* "Fuels Gasification," Advances in Chemistry Series, Vol. 69. American Chemical Society Publications, 1967.

Chapter XI

METHANOL PRODUCTION FROM ORGANIC WASTE

G. Haider
DEPARTMENT OF FUELS ENGINEERING
UNIVERSITY OF UTAH
SALT LAKE CITY, UTAH

I. INTRODUCTION

One alternative in the efforts to utilize available resources, including waste, more effectively is the transformation of these residues into marketable fuel forms. These fuel forms are synthetic gasoline or other liquid petroleum products, synthetic natural gas, and synthetic liquid fuels such as methanol. The choice of the processed form among these alternatives depends on market, overall costs, including the cost of emission control, and the ease with which this fuel form fits into the existing network of supply and distribution. These are user-oriented issues. The technical and economic evaluations, based on these criteria, indicate that methanol is a logical candidate [1].

A.   Chemical Uses

Major methanol markets have been described by Blackford [2]. The largest end use of methanol is in formaldehyde. Other chemicals derived from methanol include dimethyl terephthalate, methyl halides, methyl amines, methyl methacrylate, ethylene glycol, methyl esters, and acetic acid. Methanol is also used as solvent inhibitor for formaldehyde.

Patents and publications have been issued for many new or improved direct uses of methanol. Among these are fuel antiicer [3], fuel cell component [4], fuel additive for rocket, jet, and combustion engines [5], purification of coal, coal tar, and gaseous and liquid hydrocarbons [6], in plastics for catalyst removal, separation, and telogenation [7], in metallurgy for carburization and cementation atmospheres [8], as a seed disinfectant [9], a hydraulic cement retarder [10], and a leading agent for uranium ore [11]. Countless other patents have been issued covering methanol derivatives useful in a broad spectrum of industry [12].

B.   Fuel Potential

Methanol for direct combustion would compete with gasoline, low sulfur fuel oil, solvent-refined coal, or coal with stack gas cleanup. Methanol would also compete with fuels from other local sources.

1.   Methanol as a Transportation Fuel*   The potential of methanol as fuel for automobiles has been recognized for some time [14–19]. The interest has renewed due to Clean Air Act regulation. Methanol is a fairly volatile liquid. It has a lower volatility, a little higher density, and about half the heating value per gallon as compared to isooctane which is 100-octane gasoline reference fuel. Some of the properties of methanol are compared with isooctane and gasoline and are shown in Table I.

An appreciable experience exists with automobile propulsion systems. Methanol is used, preferentially, in high performance racing cars because it is safe and has high octane ratings.

---

* Adapted with permission from Mills and Harvey [13]. Copyright by the American Chemical Society.

TABLE I[a]

Formula	Methanol CH_3OH	Isooctane C_8H_{18}	Gasoline $C_4H_{10}C_{12}H_{26}$
Boiling point (°F)	148	211	100–400
Freezing point (°F)	−144	−161	−100
Latent heat of vaporization at the boiling point (BTU/lb)	474	117	116
Liquid density (lb/gal)	6.6	5.8	6.2
Heat of combustion			
(liquid–fuel–liquid H_2O) (BTU/lb)	9.756	26,556	20.260
(BTU/gal)	64,390	116,000	124,800
Octane No. (research)	106	100	91
Octane No. (motor)	92	100	84

[a] Reprinted with permission from G.A. Mills and B.M. Harney, *Chem. Technol.* **4**(1), 26–31 (1974). Copyright by the American Chemical Society.

Methanol alone, and as a blend with gasoline, has been tested. Adelman *et al.* [18] conducted extensive testing of methanol in a modified six-cylinder engine. The modifications of the engine were aimed at utilizing methanol fuel and minimizing auto emissions. The carburetor was rejetted to account for the fact that the stoichiometric air/fuel ratio for methanol is less than half that of gasoline and an exhaust-heated intake manifold was provided to take care of the higher latent heat of vaporization as compared to gasoline. For pollution control purposes, a catalytic muffler coupled with air injector after combustion was installed to decrease hydrocarbon and carbon monoxide emissions. The vacuum spark advance was disconnected to reduce NO_x emissions.

The results of these tests were encouraging. A methanol-fueled car with a single catalytic muffler had the exhaust emissions which were well below the 1975–1976 federal standards. Adelman and his associates also concluded that methanol emissions have less smog-producing reactivity than emissions from gasoline. Fuel consumption was tested during this test series. As expected, methanol was found to give approximately one-half the mileage of gasoline.

The use of 100% methanol has the potential of a much larger consumption as a gasoline replacement than as a blended product. However, a methanol–gasoline blend has the significant advantage of not requiring major changes in storage and distribution facilities.

2. *Methanol as a Utility Fuel* [20] Methanol can be fired in conventional boilers. Vulcan Cincinnati and New Orleans Public Service carried out demonstration firing tests which were successful. With methanol fuel,

NO_x emissions were lower than with other fuels. However, Robert Reed, a combustion expert, believes that the most important factors in a boiler are the operating conditions, such as temperature and residence time, and not the nature of the fuel [21].

Capital requirements for the two fuels, natural gas and methanol fuel, should be essentially identical for new boiler installations. However, the solvent nature of methanol would necessitate independent storage, piping, pumps, etc., to segregate methanol from any conventional fuel–oil system. The low viscosity of methanol would, in most cases, require tighter standards on valve packings, a separate feed pump, and burner orifices.

Overall thermal efficiency of boilers should be unaffected by a switch from natural gas to methanol. Although methanol has a high latent heat of vaporization (about 500 BTU/lb, or equivalent to 5% of the gross heating value) the overall boiler efficiency can be improved by using the waste heat to preheat and vaporize the fuel prior to combustion.

3. Methanol—Fuel for Fuel Cells [22]   Methanol has a unique characteristic not found in gasoline or heavier petroleum-based fuels. This characteristic is its suitability for fuel cell use. Next to hydrogen, methanol has received the most attention for potential use as a fuel for fuel cells. It has a high theoretical potential (1.2 V) and a high energy density (2.75 kWh/lb fuel).

C.   Feedstocks for Production

Methanol is produced from synthesis gas, a mixture of carbon monoxide and hydrogen. This feedstock has come, traditionally, from the pyrolysis of wood. More recently it has been produced by natural gas re-formation.

Synthesis gas can be produced from municipal solid waste, as was discussed in Chapter II. It can also be produced from wood waste or cropwaste, as demonstrated in Chapters VII–IX.

II.   METHANOL PRODUCTION CONSIDERATIONS

Production considerations include the thermodynamics of the methanol synthesis, the use of catalysts to improve production, and the kinetics of the reactions.

A.   Thermodynamics

1. Methanol Synthesis from Carbon Monoxide and Hydrogen

a. Reaction Equilibrium for Methanol Synthesis [23]   The synthesis of methanol may be represented by the equation

$$CO + 2H_2 = CH_3OH, \qquad \Delta H_{298} = -21,685 \text{ cal} \qquad (1)$$

The equilibrium constant K_p^0 of this reaction can be calculated by the equation

$$K^0_p = \frac{(N_{CH_3OH})}{(N_{CO})(N_{H_2})^2} \cdot \frac{1}{p^2} \cdot K_\gamma \tag{2}$$

where N, p, and K_γ are the mole fraction of a component, pressure of a system, and fugacity ratio in methanol synthesis, respectively. The values of the fugacity ratio of this reaction have been calculated and plotted by Ewell [24] as shown in Fig. 1.

The theoretical values of K_p^0, from Eq. (2), determine the upper limit of methanol concentration in the effluent mixture from the synthesis converter for a given set of reaction conditions such as temperature, pressure, and feed gas composition.

Some of the K_p^0 values reported in the literature were plotted versus temperature and a curve representing the average values was drawn. This plot is shown in Fig. 2.

Chezendnichenko [25] developed the following numerical equation to calculate the values of K_p^0:

$$\log K_p^0 = 3.971 T^{-1} - 7.492 \log T + 1.77 \times 10^{-3}T$$
$$- 3.11 \times 10^{-8}T^2 + 9.218 \tag{3}$$

The values calculated by using Eq. (3) were also plotted in Fig. 2. These values coincide with the reported average values in Fig. 2.

Other investigators calculated the values of K_p^0 for methanol synthesis. They used equations similar to Eq. (3), but with slightly different coeffi-

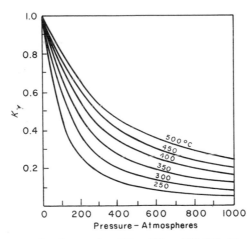

Fig. 1. Values of K_γ for the reaction $CO + 2H_2 + CH_3OH$. (Reprinted with permission from Ewell [24]. Copyright by the American Chemical Society.)

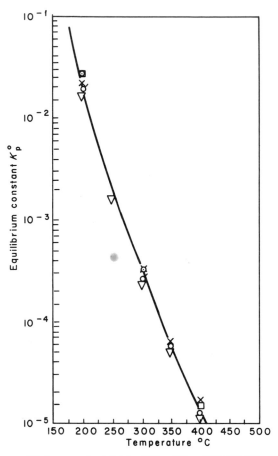

Fig. 2.   Reaction equilibrium constant K_p^0 for $CO + 2H_2 \rightleftarrows CH_3OH$. (From Strelzoff [23].)

cients of the various terms and a slightly different integration constant. Their calculated values were different because they used different values of thermodynamic properties, which have been revised from time to time as the laboratory methods for determining these values were improved.

  b.   Effect of Temperature, Pressure, and Inlet Gas Composition on the Equilibrium Methanol Concentration in the Effluent Mixture [23]   The equilibrium concentration of methanol, N_{CH_3OH}, can be determined from Eq. (2), which can be written as

$$N_{CH_3OH} = \frac{K_p^0(N_{CO})(N_{H_2})^2}{K_\gamma} \cdot p^2 \tag{4}$$

Temperature, pressure, and inlet gas composition are the basis of numerical values of N_{CO} and N_{H_2} in Eq. (4). These conditions, therefore, influence the concentration of methanol in the effluent mixture.

An increase in temperature decreases the methanol concentration in the effluent mixture. This result is brought about by two factors. First, K_γ increases with temperature as shown in Fig. 1. Second, $K_p{}^0$ decreases with temperature as indicated in Fig. 2. The overall effect of temperature on equilibrium methanol concentration is given in Table II.

Equation (4) shows that equilibrium methanol content in the effluent mixture increases with pressure to the second power. The influence of pressure is shown in Table III.

The effect of the hydrogen/carbon monoxide molar ratio on methanol content of effluent gas is given in Table IV. According to Eq. (4), the stoichiometric mixture of reactants, a hydrogen/carbon monoxide ratio of 2:1, gives the highest methanol content at equilibrium.

However, the stoichiometric mixture is not always used in actual plant operation. A ratio as high as 6:1 is often employed in the cycle gas. The excess hydrogen helps in dispersing the heat generated during the reaction. It also increases the yield of methanol based on consumption of carbon monoxide, although the conversion per pass decreases on the basis of total gas charged. The higher ratio also keeps down the side reaction. This is illustrated in Table IV.

c. *Side Reactions* [12]   Theoretically, carbon monoxide and hydrogen can react to form a series of products other than methanol. Some of these reactions are given in Table V.

Many side reactions are formed thermodynamically over the methanol synthesis reaction at all temperatures and pressures. The undesirable side

TABLE II

Effect of Temperature on Equilibrium Methanol Concentration[a]

Pressure (atm)	Temperature (°C)					
	240	280	300	340	380	400
50	20.0	13.0	8.7	2.00	0.94	0.57
100	31.7	25.7	20.4	9.95	3.05	2.43
150	32.8	30.1	27.0	17.3	8.47	5.36
200	36.1	31.8	30.1	23.0	15.3	9.44
300	33.3	32.8	32.1	28.6	21.4	17.0

[a] Mole percent in a reaction mixture from an inlet mixture with a hydrogen/carbon monoxide ratio of 4:1. (From Mikhailova *et al.* [25a].)

TABLE III

Effect of Pressure on Equilibrium Concentration of Reactants in Methanol Synthesis Reaction[a]

Pressure (atm)	x	Mole %		
		CO	H_2	CH_3OH
10	0.011	33.2	66.5	0.36
25	0.064	32.6	65.1	2.2
50	0.211	30.6	61.2	8.2
100	0.491	25.2	50.5	24.3
200	0.740	17.1	31.2	48.7
300	0.831	12.6	25.1	62.3
400	0.874	10.1	20.1	69.7
600	0.923	6.7	13.3	79.9
800	0.942	5.2	10.4	84.5
1000	0.951	4.2	8.4	87.2

[a] Mole percent in a reaction mixture from an initial hydrogen/carbon monoxide ratio of 2:1 at a temperature of 300°C. (From Comings [25b].)

TABLE IV

Effect of Inlet Gas Mixtures of Various Hydrogen/Carbon Monoxide Molar Ratios on Equilibrium Methanol Concentration[a]

Pressure (atm)	H_2 CO molar ratio					
	1:1	4:1	2:1	1:1	1:2	1:4
50	1.5	2.2	2.3	2.2	1.5	0.7
75	2.5	4.1	5.1	4.1	2.5	1.2
100	3.9	6.7	8.4	6.7	3.0	1.8
125	5.2	9.4	12.1	9.5	5.4	2.5
150	6.6	12.2	15.9	12.3	6.8	3.3
175	7.5	14.6	19.7	15.2	8.2	3.8
200	8.3	17.4	23.4	18.0	9.6	4.4
250	9.0	21.4	30.8	22.2	11.9	5.7
300	10.7	24.2	37.8	26.5	13.8	6.8
350	11.3	26.2	44.6	29.8	15.5	7.5
400	11.6	27.8	51.4	32.3	16.8	7.9
450	11.8	29.4	57.6	34.5	17.7	8.4
500	12.0	30.7	63.5	36.5	18.5	8.9

[a] Mole percent in a reaction mixture at 350°C. (From Mikhailova *et al.* [25a].)

TABLE V

Typical Side Reactions In Methanol Synthesis from Carbon Monoxide and Hydrogen[a]

Side reactions		ΔF_{298}
1.	$CO + 2H_2 \rightleftharpoons CH_3OH$	$-5{,}880$
2.	$CO + 3H_2 \rightleftharpoons CH_4 + H_2O$	$-33{,}970$
3.	$2CO + 5H_2 \rightleftharpoons C_2H_6 + 2H_2O$	$-51{,}520$
4.	$3CO + 7H_2 \rightleftharpoons C_3H_8 + 3H_2O$	$-71{,}095$
5.	$2CO + 4H_2 \rightleftharpoons CH_3OCH_3 + H_2O$	$-16{,}320$
6.	$2CO + 4H_2 \rightleftharpoons C_2H_5OH + H_2O$	-29.320
7.	$3CO + 6H_2 \rightleftharpoons C_3H_7OH + H_2O$	$-53{,}000$
8.	$2CO + 2H_2 \rightleftharpoons CH_4 + CO_2$	$-40{,}780$
9.	$3CO + 3H_2 \rightleftharpoons C_2H_5OH + CO_2$	$-36{,}140$
10.	$2CO \rightleftharpoons C + CO_2$	$-28{,}640$

[a] From Woodward [12]. Reprinted by permission of John Wiley & Sons, Inc.

reactions can be reduced to a minimum by using a selective catalyst for the formation of methanol and by choosing a set of appropriate reaction conditions.

2. Methanol Synthesis from Carbon Dioxide and Hydrogen

a. Reaction Equilibrium for Methanol Synthesis In actual practice, synthesis gas from any source is a mixture of carbon monoxide and carbon dioxide. A part of carbon dioxide is converted to methanol along with carbon monoxide. The conversion of carbon dioxide to methanol takes place according to either the one-step reaction

$$CO_2 + 3H_2 = CH_3OH + H_2O, \quad \Delta H_{298} = -11{,}830 \text{ cal} \tag{5}$$

or a two-step reaction

$$CO_2 + H_2 = CO + H_2O, \quad \Delta H_{298} = 9855 \text{ cal}$$
$$CO + H_2 \times CH_3OH, \quad \Delta H_{298} = -21.685 \text{ cal} \tag{6}$$

The values of the calculated equilibrium constant for intermediate and overall reactions for methanol synthesis from carbon dioxide are shown in Table VI. These values decrease as the temperature increases. Pressure has a favorable effect on conversion [12].

The theoretical equilibrium composition of synthesis reaction product depends on the mechanism of formation of methanol from carbon dioxide. Sufficient data are not available to draw any conclusion regarding the

TABLE VI

*Equilibrium Constant at 298 and 600 K for the Formation of
Methanol from Carbon Dioxide and Hydrogen*[a]

Reaction	K_{298}	K_{600}
$CO_2 + H_2 \rightleftharpoons CO + H_2O$	1.00×10^{-5}	3.64×10^{-2}
$CO + 2H_2 \rightleftharpoons CH_3OH$	2.09×10^{4}	9.18×10^{-5}
$CO_2 + 3H_2 \rightleftharpoons CH_3OH + H_2O$	2.10×10^{-1}	3.34×10^{-6}

[a] From Woodward [12]. Reprinted by permission of John Wiley
& Sons, Inc.

mechanism. Also, kinetics of methanol synthesis from a mixture of carbon monoxide and carbon dioxide involves five reactants and two or three equivalent constants.

Owing to these uncertainties and complicated computations, the activity tests for the catalysts are generally conducted with mixtures of carbon monoxide and hydrogen containing no carbon dioxide to evaluate the activity constants. These constants are subsequently used to calculate the rate of formation of methanol. The computation is simplified by the use of this rate of formation for inlet gas mixtures containing both carbon monoxide and carbon dioxide with the assumption that the two-step reaction mechanism is valid and that carbon monoxide produced in the first step is taken into account. However, the effect of carbon dioxide is lumped together with that of the residual carbon monoxide. This is not strictly correct; however, the plants designed on these bases have not given any trouble as far as material and heat balances are concerned [23].

Thus, in practice, a $H_2/(2CO + CO_2)$ ratio in place of a H_2/CO ratio is used. The conversion is based on both carbon monoxide plus carbon dioxide and not on carbon monoxide alone. The heat of reaction is smaller in the case of synthesis from carbon dioxide and hydrogen. This is advantageous because less heat has to be removed from the catalyst bed [23].

The disadvantages of carbon dioxide in methanol synthesis are that 50% more hydrogen is consumed than by using carbon monoxide, and consequently there is a higher compression cost. But these disadvantages are overrun by elimination of the necessity for the removal of carbon dioxide from synthesis gas [23].

b. Side Reactions [12]   The carbon dioxide and hydrogen may react to form carbon monoxide and water. Theoretically, methanation can take place according to the reaction

$$CO_2 + 4H_2 = CH_4 + H_2O \qquad (7)$$

This reaction is favored thermodynamically over the methanol synthesis. However, the heat generated during the reaction is less in the case of carbon dioxide, and this minimizes the formation of methane. Also, pressure has a favorable effect on the formation of methanol.

Above all, the selectivity of the catalyst employed in the reactor plays an important role in suppressing undesirable reactions.

3. *Heat Balance* [23]   The heat of reaction for methanol synthesis from carbon monoxide and hydrogen is 21,700 cal/gm mole at 25°C and 1 atm. It increases with the reaction temperature. Thomas and Portalski [26] developed the following equation to calculate the heat of reaction at higher temperatures and 1 atm pressure:

$$H_{T,P^0} = -17,920 - 15.84T + 1.142 \times 10^{-2}T^2 - 2.699 \times 10^{-6}T^3 \quad (8)$$

For heat of reaction at pressures higher than 1 atm the same authors developed an equation based on the critical constants of the reactants and Berthlot's equation [27]. The equation is as follows:

$$H_{T,P^0} = H_{T,P} - 0.5411P - 3.255 \times 10^6 T^{-2}P \quad (9)$$

The plot of $H_{T,P}$ values calculated using Eq. (9) is shown in Fig. 3.

The heat of reaction increases appreciably with increase in pressure, especially above 300°C. For example, the heat of reaction is approximately

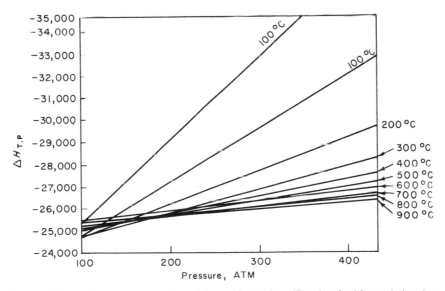

Fig. 3.   Effect of pressure on methanol heat of reaction. (Reprinted with permission from Thomas and Portalski [26]. Copyright by the American Chemical Society.)

27,200 cal/gm mole at 350°C and 350 atm versus 24020 cal/gm mole at 350°C and 1 atm. Heat of reaction would be in the order of 639 million cal or 2.53 billion BTU/day for a plant of 750 m tons of methanol per day, operated at 350°C and 350 atm. This amount of heat is equivalent to 88.5 m tons of high grade (13,000 BTU) coal per day.

The enormous amount of heat generated during the reaction is recovered and a heat balance around the plant can be worked out with very little heat loss.

B. Catalysis

From a practical standpoint, a catalyst should possess high activity, good resistance toward aging, and high selectivity. Using these criteria, the performance of different catalysts for methanol synthesis has been studied. It is practically impossible to correlate the results of different experiments and draw quantitative conclusions. The complication arises from the wide differences in the experimental procedures that were followed. Some investigators studied the synthesis of methanol directly, whereas others used the decomposition of methanol as the basis to evaluate the catalysts for methanol synthesis. Natta [28] states that particularly from the point of view of selectivity, there is a lack of similarity between the catalytic activity in the reaction of synthesis and in that of decomposition of methanol. But Frolich *et al.* [29] have concluded from their experimental data that the decomposition method of testing is reliable for studies of the activity of catalysts for synthesis of methanol from synthesis gas at high pressures.

1. Catalyst Characteristics   The characteristics of a catalyst depend on a number of factors:

(1)   amount of precipitating agent [28],
(2)   amount of promoter [30],
(3)   temperature of calcination [28],
(4)   temperature of reduction [31–33],
(5)   rate of reduction [32],
(6)   reducing agent [32,34,35],
(7)   size of the catalyst [36,37],
(8)   shape of the catalyst [38],
(9)   tableting pressure [39–41].

In spite of these circumstances, it is possible to make a qualitative comparison of catalysts that have been studied by different methods.

The studies of the catalysts have been reviewed by Natta [28]. This has been summarized along with the results of the work reported by Frolich and co-workers.

2. Catalysts to Consider

a. Zinc Oxide [28] Pure ZnO is not used in commercial plants. However, it is present in the majority of the widely used catalysts. Therefore, the knowledge of catalytic performance of pure ZnO is important for the interpretation of the mechanism of catalysis with mixed catalysts.

Zinc oxide is one of the most selective catalysts for the synthesis of methanol.

The behavior of a catalyst depends on the method of preparation of the catalyst. Zinc oxide can be prepared by the combustion of metallic zinc, by calcination of precipitated zinc hydroxide or carbonate, or by thermal decomposition of molten salts such as zinc acetate.

The catalytic activity of ZnO seems to depend on the anion of the source compound for precipitation [42]. For example, catalysts obtained from $Zn(NO_3)_2$ are more active than those obtained from $ZnCl_2$ or $ZnSo_4$.

The activity of ZnO prepared by thermal decomposition of zinc compounds depends on the temperature of preparation.

The catalytic activity is influenced more by the crystalline size than by the particle size [43]. It has been confirmed that smaller crystalline size improves the activity. The crystal size depends on the source material of ZnO catalyst.

The aging of ZnO catalyst depends on the source material used to prepare ZnO. The zinc oxides obtained from smithsonite (mineral $ZnCo_3$) and zinc acetate are stable and have been used to study the action of these catalysts. The catalysts from smithsonite and zinc acetate showed activation energies of 27 and 30 kcal/mole, respectively [43].

b. Copper The decomposition of methanol was studied by Sabatier and Senderens [44] and Lormand [45] over the oxides of metals, such as vanadium, chromium, and zinc, and copper powder. The results indicate that powdered copper is the most active catalyst for methanol synthesis. However, this conclusion is contradicted by experience, and pure copper gives very poor results. On the other hand, oxides of vanadium, chromium, and zinc resulted in the formation of methanol as expected.

Pure copper oxide shows a weak catalytic activity. Copper oxide reduces itself to metallic copper which then crystallizes readily. Reduced copper oxide gives catalytic activity which depends on the temperature of reduction [43,46,47].

c. Chromium Oxide [28] The different types of the more or less hydrated chromium oxide are very poor catalysts for the synthesis of methanol. The chromium oxide obtained by the decomposition of $Cr(OH)_3$

shows the highest activity in this group [48]. The chromium hydroxide is obtained from solutions of $Cr(NO_3)_3$ and NH_3. This catalyst has an activity which is considerably higher than that of the chromium oxide obtained from chromium oxalate.

Other metallic oxides such as MnO and MgO have been proposed for the preparation of mixed catalysts. These oxides do not show any remarkable activity for the synthesis of methanol by themselves.

   d.   Mixed Catalysts   The catalysts of actual industrial importance for the synthesis of methanol are composed of mixtures of two or more oxides.

C.   Kinetics

1.   Rate Equation [28]   The kinetics of methanol synthesis was studied by Natta and co-workers [49]. They concluded that the kinetic equation based on the hypothesis that the surface reaction is trimolecular fit the isothermal rate curves. The equation is

$$ r = \frac{\gamma_{CO}p_{CO}\gamma^2_{H_2}p^2_{H_2} - (\gamma_{CH_3OH}p_{CH_3OH}/K_{eq})}{A + B\gamma_0 p_{CO} + C\gamma_{H_2}p_{H_2} + D\gamma_{CH_3OH}p_{CH_3OH}} \tag{10} $$

Where p and γ are the partial pressure and fugacity coefficient, respectively, and A, B, C, and D are positive constants which are functions of temperature only.

The trimolecular surface reaction is represented by the scheme

$$ Co^* + 2H_2^* = CH_3OH^* + 2L $$

Where the asterisk indicates the absorbed species on the active site L of the catalyst.

For the initial rate of reaction r_0 when the methanol concentration is zero, Eq. (10) is simplified to

$$ A + B\gamma_{CO}p_{CO} = C\gamma_{H_2}p_{H_2} = \frac{3(\gamma_{CO}p_{CO}\gamma^2_{H_2}p^2_{H_2})^{1/2}}{r_0} \tag{11} $$

Equation (11) can be solved for three unknowns by substituting the measured initial rates of reaction in three experiments performed at the same temperature but at different pressures and synthesis gas composition. This is repeated at different temperatures, and the values of A, B, and C are determined as functions of temperature. To evaluate constant D, the determined values of A, B, and C are substituted into Eq. (10) which is solved by introducing the experimental values of r resulting from different values of the time factor and with the experimental values of the partial pressures of CO, H_2, and CH_3OH.

The values of the constants A, B, C, and D in the kinetic equation (10) are established for a particular catalyst. Then it becomes possible to predict the reaction rate at any set of operating conditions (temperature, pressure, and gas composition) within the limit of the applicability of the kinetic equation.

2. *Application to Design* [23]   The activity constants of the catalyst are determined experimentally in the laboratory with a stoichiometric mixture of carbon monoxide and hydrogen under strict isothermal conditions as described in the preceding section.

The composition of inlet gas is a part of the design basis for methanol synthesis. The stoichiometric ratio of carbon monoxide to hydrogen gives the highest methanol content in the effluent gas at equilibrium. However, hydrogen is consumed in side reactions. The inerts are always present in the synthesis gas. The concentration of the inerts is maintained within limits by purging. Therefore, hydrogen is lost in the purge gas. The carbon dioxide is not completely removed from the synthesis gas because of economic reasons. Extra hydrogen is consumed in reducing carbon dioxide to carbon monoxide. At high pressures, the rate of the synthesis reaction is high and the heat of reaction builds up rapidly. Excess hydrogen is an effective coolant and, furthermore, it suppresses undesirable reactions. Aforementioned reasons necessitate the choice of inlet gas composition which results in the best economy in both materials and energy.

The percentage of conversion of carbon monoxide and carbon dioxide to methanol is another important design consideration. This sets up the cutoff point in the path of methane reaction toward equilibrium so as to avoid the slower rate of reaction near the point of equilibrium. After establishing the inlet gas composition, the design percentage of conversion is set at about one-half the theoretical equilibrium conversion. For example, an inlet gas containing 15% carbon monoxide and carbon dioxide under a set of certain conditions gives an equilibrium concentration of 60%. Then 30% conversion is used as a design basis. These bases can be used to calculate the partial pressures of carbon monoxide, hydrogen, and methanol. The rate of formation of methanol can be calculated by substituting these values in Eq. (10). For this rate of methanol formation, the inlet gas space velocity can be calculated. This, then, gives all the necessary data for designing a commercial methanol plant as far as material balance is concerned.

The design of an optimal industrial reactor which incorporates the stoichiometric, thermodynamic, kinetic, and heat transfer considerations has been reported by many workers [50–56].

III. MANUFACTURING PROCESSES OF METHANOL

Major process and operation units for the manufacture of methanol are shown in Fig. 4. Methanol synthesis processes are classified on the basis of operating pressure. These processes are classified as (1) high, (b) low, and (c) medium pressure processes.

A. High Pressure Process

The synthesis gas is compressed to about 765 psig in a two-case centrifugal compressor, directly coupled to a steam turbine with no gear. The source [23]. The synthesis gas designates the makeup gas only, and should not be confused with the feed gas, which includes recycle gas, to the converter. The feed gas is compressed to about 5000 psig.

In moderate-sized plants, capacity less than 700 tons/day, the compression unit consists of a set of centrifugal and reciprocating piston type of compressors in series. The centrifugal compressor is used in the initial stages only because it has low efficiency as compared to the reciprocating piston type of compressor at a pressure higher than 3500 psig in a moderate capacity plant. However, the efficiency of a centrifugal

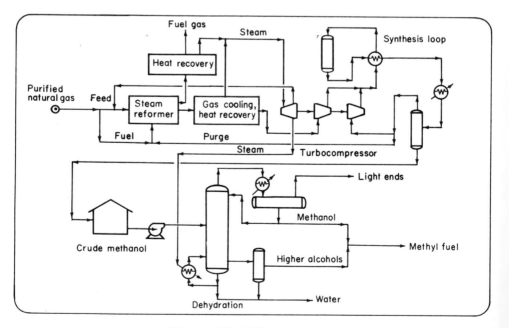

Fig. 4.   Methyl fuel production.

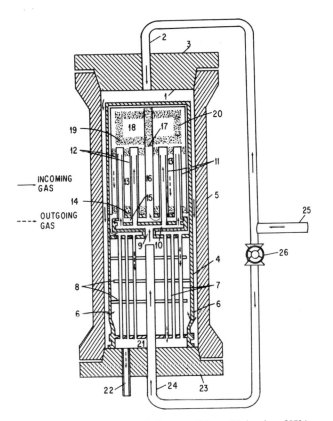

Fig. 5. Methanol reactor tubular type. (From Richardson [57].)

compressor increases with plant capacity, and all centrifugal compression units become feasible for a 700-ton/day-capacity plant.

In a methanol plant of 800 tons/day capacity, the synthesis gas is compressed by a steam-turbine-driven, five-case centrifugal compressor to a discharge pressure of 5200 psig. The recycle gas from the synthesis loop is introduced into the last case, before the final stage or next to the final stage, of this compressor. It combines with the makeup gas and the mixture is passed on through the last compression stage to the converter.

The recycle gas contains an excess amount of hydrogen, often as much as 100% over the hydrogen content in a stoichiometric mixture. But the mixture of recycle and makeup gases has a $H_2/(2CO + 3CO_2)$ ratio as high as 2.2. This gas enters the converter at a space rate of 25,000 to 35,000 hr^{-1}.

The design of the converter is an important feature, because fairly large quantities of heat are generated from the exothermic methanol synthesis reaction.

In a moderate-sized plant, the converter consists of two catalyst beds in series, as shown in Fig. 5. The first stage bed operates under adiabatic conditions. The second stage bed temperature is controlled by circulating cold feed gas through the tubes distributed in the catalyst bed. The heat in gases from the second stage catalyst bed is utilized to preheat the feed gas. The heat exchanger for this purpose is built into the converter. The temperature profile along the catalyst bed shows a temperature of 330 to 340°C at the feed gas inlet of the first bed and 380 to 390°C at the gas outlet of the second bed.

For large-sized plants, the synthesis converter design is based on the quench principle. Such a converter is shown in Fig. 6. It consists of four or five beds of catalyst. Each bed operates under adiabatic conditions. The catalyst bed temperature is controlled by injecting cold feed gas at the top of each bed. The flow rate in these jets is automatically controlled to maintain the optimum temperature in all parts of all the catalyst beds in the converter. In some designs, temperature control is accomplished by installing water-cooled coils between the catalyst beds. The thickness of these beds increases in the direction of gas flow. This arrangement compensates for the decreasing reaction rate as the reaction approaches equilibrium.

The effluent gas from the converter at a space rate of about 30,000 hr⁻¹ contains about 5 mole % methanol. This corresponds to a conversion of 29 to 30% per pass based on carbon oxides in the feed. The yield is 95 to 96%

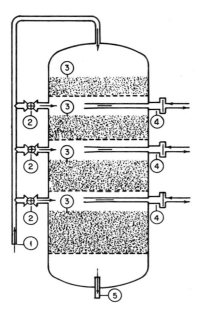

Fig. 6. Quench type of methanol synthesis converter with cooling coil. (From L'Homme [51].) 1, Inlet for the combined synthesis gas; 2, quenching gas control valves; 3, catalyst beds; 4, cooling coils (optional); 5, exit of converter gas.

based on the consumption of carbon oxides. The production rate is in the range 0.05 to 0.08 short tons/ft³ catalyst-hr, corresponding to the percentage of inerts of 30 to 10%, respectively, in the feed gas.

The effluent from the converter is cooled to condense methanol, and the condensate flows to a gas–liquid separator. The gas is recycled. The liquid from the separator is reduced in pressure to 40 psig as it flows through the letdown valve. The liquid and flashed gases and vapors flow to a flash tank. The liquid, containing 75 to 90% methanol, is the crude product which is purified further. The flashed gases and vapors are scrubbed with water to recover methanol vapors. The gases escaping from the top of the scrubber are accounted for as part of the purge gas.

The crude methanol contains impurities such as acetone, aldehyde, amines, dimethyl ether, ethanol, and higher alcohols. These impurities total less than 3.0% in the crude methanol.

The crude methanol is treated with potassium permanganate which converts the aldehydes to organic acids. The manganese precipitates are removed by filtration. The filtrate is neutralized with sodium hydroxide solution and then subjected to fractionation, for which two fractionating columns are used. In the first column, acetone is extracted with water. In the second column, ethanol, higher alcohols, and other impurities are removed from the respective fractionating zones to give commercial grade of 99.5% methanol. The bottom stream of the second column is pure water, which is recycled and used in the scrubber and first fractionating column as described earlier.

B. Low Pressure Process

The manufacturing scheme of this process is similar to that described for the high pressure process [23]. Changes in the processing and operating procedures in the low pressure process are as follows.

Synthesis gas is treated to remove sulfur thoroughly to avoid poisoning of the catalyst. It is not essential to remove carbon dioxide completely from the synthesis gas. The presence of carbon dioxide in the feed gas is beneficial in more than one way. First, the presence of as little as 0.31% carbon dioxide in feed gas can double the percentage of conversion of carbon oxides to methanol. Second, a substantial amount of carbon dioxide in the feed gas improves the life of the catalyst. For example, with 15% carbon monoxide plus 10% carbon dioxide in the feed gas, the activity of a copper–zinc–chromium catalyst is sustained without deterioration even after six months continuous operation, whereas without carbon dioxide, or with insufficient carbon dioxide, the catalyst would lose 6 to 8% of its activity in a few days.

The synthesis gas is compressed to about 765 psig in a two-case centrifugal compressor, directly coupled to a steam turbine with no gear. The recycle gas at 690 psig is compressed to 765 psig in a separate circulating centrifugal compressor. Synthesis gas is combined with recycle gas and fed to the converter.

The feed gas contains hydrogen and carbon oxides in the stoichiometric ratio for methanol synthesis reaction with a small amount of excess hydrogen. The carbon monoxide-to-carbon dioxide molar ratio varies from 20:1 to 1:2. However, the preferred carbon monoxide-to-carbon dioxide ratio is in the range of 3:2 to 4:3.

The converter is of the quenched-bed type in the low pressure process. The temperature of the catalyst is controlled by quenching with cold feed gas as described in Section III.A. Cooling coils between catalyst beds are not used. Operating temperature and pressure are 270°C and 750 psig, respectively. The inlet space velocity is in the range of 7000 to 20,000 hr^{-1}, preferably 9000 to 11,000 hr^{-1}. The catalyst is copper based. The low pressure has an adverse effect on the methanol synthesis equilibrium, but this disadvantage is partially compensated by relatively lower operating temperature.

The effluent from the synthesis converter contains 2.5 mole % methanol at a space rate of 10,000 hr^{-1}. This corresponds to a production rate of 0.0119 short ton/ft^3 catalyst-hr. Low rate of production is undesirable, but is helpful in exothermic methanol synthesis reactions since it generates less heat per unit volume of the catalyst bed per hour. Consequently, the installation of cooling coils between the catalyst bed becomes unnecessary, and the need for excess hydrogen in the loop gas serving as coolant is eliminated.

The low pressure of the process does not mean a saving of power consumed in circulating the loop gas based on kilowatt-hours per tons of methanol produced, because although the number of moles of gas circulated per pass is reduced by nearly 40%, the production per pass is also reduced as much.

The crude methanol contains fewer impurities, because fewer side reactions take place in the converter due to low temperature and high selectivity of the copper-based catalyst. The commercial grade methanol is recovered from crude methanol by a method similar to the one used in the high pressure process, except that the pretreatment of crude methanol is not necessary.

The low pressure process is very attractive for medium-sized methanol plants because of its simple compression equipment and highly selective catalyst.

C. Medium Pressure Process

The operating pressure in this process is about 1475 psig [58]. The medium and low pressure processes resemble one another in many ways. For example, product purity, feedstock desulfurization severity, and maintenance and manpower requirements are almost identical. However, there are some major differences, in addition to the new catalyst, relating mainly to equipment size.

The per pass conversion of carbon oxides to methanol is higher. The effluent gas from the cold shot converter contains about 5 mole % methanol. This results in a reduction of interchanger heat load and, hence, the size. The circulation rate is less and, therefore, the volume of piping work is smaller. The yield of methanol on the basis of feed carbon is higher. Higher conversion per pass and higher yield require smaller converter and catalyst volume, as well as smaller diameter of loop piping.

However, in the medium pressure process, the size of the compressor is larger because of the increased process pressure. It also requires higher design pressure in the loop equipment and piping.

D. The Economics of Methanol Production

The production cost of methanol depends primarily on the raw materials chosen, the type of process used, and the size of the plant.

The sales price of methanol for an approximately 20% rate of return on investment before tax in new plants with different feedstocks, types of processes, and plant capacities was evaluated by Hadley et al. [59].

Medium-sized plants use the low pressure process. The high pressure process has an advantage over the low pressure process when the capacity of the plant is higher than the critical size of the plant. For the very large plants, 20,000 to 30,000 tons/day of methanol fuel production, high pressure technology will again be competitive [20].

In brief, for a proper choice of a process for a particular project, there are certain factors that must be taken into account along with the data presented in this section. These factors include local marketing conditions, the setup of existing facilities, available raw materials, and other economic aspects of each individual plant [23].

IV. CONCLUSION

Methanol does offer a highly attractive alternative when producing fuels from waste. It can be used in the production of chemicals, as a transportation fuel, or as a utility fuel. It is storable, transportable, and clean burning.

The production chemistry for methanol is well known and demonstrated. Commercial production systems of high, low, and medium pressure are available, and their economics are documented.

Thus, organic residues can be converted into synthesis gas and then methanol. This offers a highly attractive route for the production of a versatile waste-based fuel.

REFERENCES

1. M. Steinberg, F. J. Salzano, M. Beller, and B. Manowitz, BNL 17800. Brookhaven National Laboratory, New York, 1973.
2. J. L. Blackford, in "Chemical Economics Handbook," pp. 674.5021A–674.5030G. Stanford Research Institute, Stanford, California, 1971.
3. S. R. Oblad and M. Hamer, U.S. Patent 3,036,902.
4. W. Vielstich, French Patent 1,371,815.
5. R. L. Heinrich, J. A. Anderson, Jr., and N. P. Peet, U.S. Patent 2,916,366.
6. P. S. Landis and H. D. Norris, U.S. Patent 3,126,331.
7. N. N. Hochgraf, R. H. Schatz, and B. R. Tegge, U.S. Patent 3,052,665.
8. E. Pouillard. *Rev. Met. (Paris)* **58**, 401–406 (1961).
9. F. Wagner, German Patent 1,045,157.
10. J. Van Heininger and G. J. Harmsen, U.S. Patent 2,816,043.
11. J. M. Kruse, U.S. Patent 3,089,885.
12. H. F. Woodward, in "Kirk-Othmer Encyclopedia of Chemical Technology" (Anthony Standen, ed.), Vol. 13, pp. 370–398. Wiley (Interscience), New York, 1967.
13. G. A. Mills and B. M. Harvey, *Chem. Technol.* **4** (1), 26–31 (1974).
14. E. S. Starkman, K. H. Newhall, and R. D. Sutton, in "Alcohols and Hydrocarbons as Motor Fuels." SP-254 Society of Automotive Engineers, Inc., New York, 1964.
15. J. S. Ninomiyá, A. Golovoy, and S. S. J. Lahana, *Air Pollut. Contr. Ass.* **20** (5), 314–317 (1970).
16. R. E. Fitch and J. D. Kilgore, Consol. Eng. Technol. Corp., PB Rep. No. 194688, 1970.
17. S. J. W. Pleeth. *J. Inst. Pet. (London)* **38** (346), 805–819 (1972).
18. H. G. Adelman, D. G. Andrews, and R. S. Devoto, Paper 720693 presented at the Society of Automotive Engineers, p. 16, National West Coast Meeting, San Francisco, California, 1972.
19. G. D. Ebersole and F. S. Manning, Paper 720692 presented at the Society of Automotive Engineers, p. 24, National West Coast Meeting, San Francisco, California, 1972.
20. B. Dutkiewicz, *Oil Gas J.* **71** (18), 166–178 (1973).
21. J. C. Davis, *Chem. Eng.* **80** (15), 48–50 1973.
22. J. R. White, C. N. Rowe, and W. J. Koehl, Paper presented at Engineering Foundation Conference, Henniker, New Hampshire, 1974.
23. S. Strelzoff, in "Methanol Technology and Economics" (G. A. Danner, ed.), AIChE Symposium Sereis, Vol. 26, No. 98, pp. 54–68, 1970.
24. R. H. Ewell, *Ind. Eng. Chem.* **32**, 149–152 (1940).
25. V. M. Chezendnichenko, Dissertation, Karpova Physico–Chemical Institute, Moscow, USSR, 1953.
25a. S. A. Mikhailova, I. P. Sidorov, and D. B. Kazarnovskaya, Dissertation, USSR State Inst. of Notrogen Industry, Moscow, 1966.
25b. E. W. Comings, "High Pressure Technology," pp. 341–366. McGraw-Hill, New York.

26. W. J. Thomas and S. Portalski, *Ind. Eng. Chem.* **50**, 967–970 (1958).
27. R. Gilmont, *in* "Encyclopedia of Chemical Technology," Vol. 7, p. 110. Wiley (Interscience), New York, 1951.
28. G. Natta, *in* "Catalysis" (P. H. Emmett, ed.), pp. 349–412. Van Nostrand-Reinhold, Princeton, New Jersey, 1955.
29. P. K. Frolich, M. R. Fenske, P. S. Taylor, and C. A. Southwich, Jr., *Ind. Eng. Chem.* **20**, 1327 (1928).
30. L. E. Sushchaya, P. G. Bondar, and V. M. Vlasenko, *Katal. Katal.* **10**, 60–64 (1973).
31. M. Kawamura and T. Irie, *Kogyo Kagaku Zasshi* **60**, 696–700 (1957).
32. T. Irie and T. Shiraishi, *Nippon Kagaku Zasshi* **80**, 107–110 (1959).
33. Y. V. Zhigailo, L. I. Shpak, T. P. Gaidei, V. I. Duchinskaya, and V. V. Raksha, *Khim. Prom.* **1**, 29–34 (1963).
34. V. M. Vlasenko, M. G. Rozenfeld, and M. T. Rosov, "Nauchn Osnovy Podbora i Proizv," pp. 160–167. Katalizatorov, Akad. Nauk USSR, Sibirsk otd., 1964.
35. A. M. Alekseev, N. P. Opdounikova, and L. P. Kirillou, *Iso. Vyssh. Uchch. Zaued, Khirn, Teknol.* **10** (2), 189–193 (1967).
36. C. F. Brown, Dissertation, Univ. of Connecticut, Storrs, Connecticut, 1968.
37. D. B. Chistozvonov, V. S. Soholevskii, L. I. Kozlov, B. I. Shteinberg, A. A. Zuev, and V. I. Korbutou, *Tr. Nauchno-Issled. Proekt. Inst. Azotn Prom. Prod. Org. Sin.* **16**, 110–118 (1972).
38. M. Trabucchi and R. E. Cunningham, *Ind. Quim.* **27** (2), 52–56 (1969).
39. V. M. Vlasenko, N. K. Lunev, M. T. Rusov, and Z. V. Sopova, *Ukr. Khim. Zh. (Russian)* **32** (4), 348–353 (1966).
40. A. I. Mal'chevskii, V. M. Vlasenko, Y. V. Zhigailo, and I. D. Karpovich, *Khim. Tekhnol. (Kiev)* **5**, 15–17 (1971).
41. D. S. Shishkov and I. P. Domhalou, *Dokl. Bolg. Akad. Nauk* **26** (1), 89–92 (1973).
42. G. F. Huttig, O. Kostclitz, and I. Feher, *Z. Anorg. Allg. Chem.* **198**, 206 (1931).
43. G. Natta and P. Corradini, *Proc. Int. Symp. Reactivity of Solids, Gothenburg, 1952*, pp. 619–632.
44. Sabatier and Senderens, *Ann. Chem. Phy.* **8** (4), 469 (1905).
45. C. Lormand, *Ind. Eng. Chem.* **17**, 430 (1925).
46. E. Audibert and A. Raineau, *Ind. Eng. Chem.* **20**, 1105 (1928).
47. V. A. Plotnikov and K. N. Ivanov, *All-Ukrain. Akad. Sci. Memo. Inst. Chem.* **1**, 127 (1934).
48. M. C. Molstad and B. F. Dodge, *Ind. Eng. Chem.* **27**, 134 (1935); V. A. Plotnikov and K. N. Ivanov, *All-Ukrain. Akad. Sci. Memo. Inst. Chem.* **1**, 127 (1934).
49. G. Natta, P. Pino, G. Mazzanti, and I. Pasquon, *Chim. Ind. (Milan)* **35**, 705–725 (1953).
50. G. A. L'Homme, *RUM, Rev. Universelle Mines* **111** (4), 121–128 (1968).
51. G. A. L'Homme, *RUM, Rev. Universelle Mines* **111** (5), 149–156 (1968).
52. I. Pasquon and M. Dente, *Ind. Genie Chim.* **101** (10), 1431–1438 (1969).
53. A. Cappelli, A. Collina, and M. Dente, *Adv. Chem. Ser.* **109**, 35–37 (1972).
54. P. Uronen, *Kem Tepllisuns* **29** (5), 345–354 (1972).
55. G. A. Efankin, Y. I. Paslasvskii, and V. V. Suzdalevich, *Khim. Tekhnol. (Kiev)* **2**, 29–33 (1972).
56. A. Capelli, A. Collina, and M. Dente, *Ind. Eng. Chem. Process Des Develop.* **11** (2), 184–190 (1972).
57. R. S. Richardson, U.S. Patent 1,909,378.
58. P. L. Rogerson, *Chem. Eng. (New York)* **80** (19), 112–113 (1973).
59. B. Hadley, W. Powers, and R. B. Stobaugh, *Hydrocarbon Process.* **49** (9), 275–280 (1970).

Chapter XII

UTILIZATION OF ENERGY FROM ORGANIC WASTES THROUGH FLUIDIZED-BED COMBUSTION

G. Ray Smithson, Jr.
ENVIRONMENTAL TECHNOLOGY PROGRAM OFFICE
BATTELLE-COLUMBUS LABORATORIES
COLUMBUS, OHIO

I. INTRODUCTION

Fluidized-bed combustion has become a widely considered method for the disposal of industrial wastes. It incorporates both waste disposal and energy recovery features. Fluidized beds have also been used for wood waste gasification, and have been considered for municipal waste-to-energy systems.

Many industrial uses have been proposed for fluidized-bed systems since this technique was proposed by C. E. Robinson about a century ago. However, it was not until the late 1920s that the first commercial fluidized-bed unit involving a gas–solids mixture and utilizing elevated temperatures was installed. This installation was the beginning of the use of fluidized-bed catalytic crackers by the petroleum refining industry. Since then fluidized-solids technology has been firmly established as a useful and very valuable industrial operation. The development of this technology for use in many industries has taken years for fruition, and new applications continue to be explored extensively.

After its adoption by the petroleum industry, fluidized-bed technology was successfully applied to many gas–solid operations in other industries where careful control of temperature and gaseous atmosphere as well as simplicity of operation was needed. Included in these applications were metallurgical processes (e.g., roasting of sulfide ores, reduction of oxide ores, chloridization), carbonization of coals, drying of heat-sensitive materials, and the selective condensation of volatile compounds. The forerunner of the use of fluidized-bed combustion systems for the disposal of carbonaceous wastes probably was the use of this technique for nuclear waste disposal. Many methods for disposing of radioactive wastes were studied by investigators for the Atomic Energy Commission in the late 1950s and various types of fluidized-bed systems were investigated extensively for this purpose.

One of the first applications of fluidized-bed technology for the incineration of carbonaceous industrial wastes was in the pulp and paper industry. Research on this application was begun at the Columbus Laboratories of Battelle Memorial Institute in the late 1950s and culminated in the erection of a Container-Copeland commercial installation at the Carthage, Indiana mill of Container Corporation of America in 1962. The initial application at Carthage continued in successful operation until pulping operations were discontinued; a second unit, at the Circleville, Ohio mill of Container Corporation of America, has been in successful operation since 1964. These two installations accelerated fluidized-bed application in the pulp and paper industry so that, at present, more than 25 commercial installations have been constructed. This technique has also been applied to the disposal of wastes from many other industrial operations and the effective utilization of the heating value of their carbonaceous constituents.

II. GENERAL DESCRIPTION OF TECHNOLOGY

Research on the fluidized-bed incineration of waste pulping liquor began with bench-scale equipment similar to that shown in Fig. 1. Following the successful completion of the bench-scale study, the experimentation was continued in pilot-scale fluidized-bed equipment. A typical pilot-scale

Fig. 1.   Bench-scale fluidized-bed system.

fluidized-bed incineration system is shown in Figs. 2 and 3. This unit was designed to conform to a geometrical configuration similar to that used in successful commercial practice for the disposal of waste organic liquids. A detailed sketch of the pilot unit is shown in Fig. 2. Figure 3 is an overhead view of the pilot plant facility, emphasizing the 13-ft overall height of the fluidized-bed unit. The inside diameter of the bed zone of the unit is 24 in.; this zone is lined with 9 in. of insulating refractory.

The orifice plate is constructed of Inconel 600. The plate contains orifice tubes located on 4-in. diamond pitch. The orifice tubes are capped to prevent the bed from draining into the combustion chamber.

The fluidized-bed unit is mounted on a windbox combustion chamber in which natural gas is burned with an excess of air to provide a hot mixture of air and combustion gases. These gases pass through the orifice plate to fluid-ize the solid material and to provide a portion of the heat required in the operation. The dust-laden exhaust gases from the unit are discharged through a 6-in. opening in the top of the unit into the primary cyclone where the greater part of the inorganic material present in the feed mixture is separated and returned to the bed through an air ejector. After the primary cyclone, the gases pass through a venturi-type scrubber and then through an entrainment separator. The mixture of solids and liquid from the entrainment separator flows into a sump which provides a water seal for the

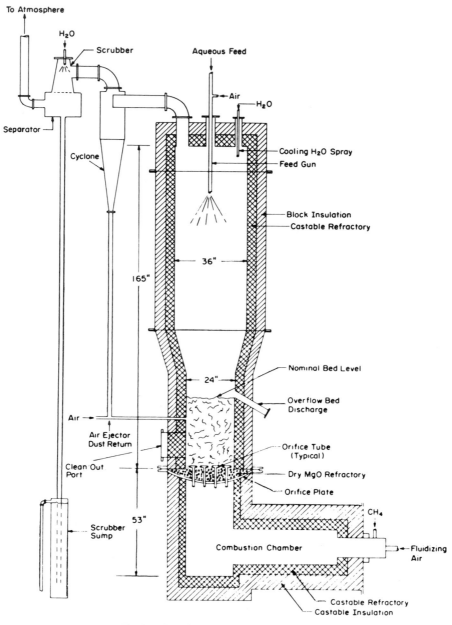

Fig. 2.   Sketch of 2-ft fluidized-bed system.

Fig. 3. Overhead view of the pilot plant facility, emphasizing the 13-ft overall height of the fluidized-bed unit.

scrubbing system. The effluent from this sump overflows to waste. The gases emerging from the entrainment separator are exhausted from the building through a 5-in. exhaust line.

Thermocouples and pressure taps are placed at strategic points in the system to measure the temperature of the inlet gases, the bed material, the freeboard space, and the off-gas streams, and the pressure differential across the entire system, the bed zone, and the dust collection system. The temperature of the bed is controlled during the operation of the unit by varying the flow of natural gas and/or the rate of introduction of the feed mix into the unit.

During operation, the feed materials are premixed in a heated tank containing a high speed agitator and then introduced into the reactor by a Moyno pump through a pneumatic feed gun. The feed gun is a single-barrel water-cooled type located in the center of the top of the fluidized-bed unit and is constructed in such a manner that the feed gun-to-bed distance can be varied. The feed mixture is dispersed into droplets of the desired size by varying the amount of air to this feed gun. The solid residue from the incineration of the feed mix is discharged from the unit through the 30-in. overflow type of discharge port periodically. This pilot-scale unit has been used in a variety of research and development programs in which the feasibility of burning wastes ranging from paint solvents to rice hulls has been investigated. In several instances these studies have led to the construction and operation of commercial systems. A few of these commercial installations are described in this chapter.

III. ENERGY UTILIZATION

Many of the troublesome organic-containing wastes generated by industry are diluted by vast quantities of water used to wash unwanted constituents from the products and to cleanse equipment and floors. The utilization of the energy contained in these dilute wastes thus is complicated by the water which carries them away. To utilize this energy source requires either the separation or concentration of the organic constituents of the wastewater.

If the organic constituents are insoluble and perhaps of a reasonably large particle size, they can be separated from the wastewater by physical means, i.e., screens, filters, flotation equipment, and the like. If, however, the organic components are soluble, more rigorous means of concentration must be employed before the energy can be utilized effectively. In some instances it is feasible to use biological treatment of organic-containing wastes to produce a sludge which can be burned. In other cases, the use of multiple-effect evaporators to concentrate the organic components has proven effective and economical. The use of evaporative procedures to concentrate waste liquors from paper pulping operations from about 10% of solids to about 35% of solids is discussed later in this chapter. This concentration is sufficient to permit the combustion of the concentrated waste liquor without the addition of supplemental fuel.

The combustion of industrial wastes for energy utilization often is complicated further by the presence of mixtures of various residual components which have relatively low melting points. This, of course, limits the temperature at which a fluidized-bed combustion system can be operated to avoid defluidization of the bed and leads to the production of exhaust gases at temperatures that may be too low for the economical recovery of its sensible heat.

Some industrial waste materials can be utilized to raise steam and to generate electricity. The inherent features of fluidized-bed systems often may be used more advantageously for the disposal of troublesome industrial wastes by utilizing their heat content to evaporate the water which they contain. When the temperature of the exhaust gases is sufficiently high, the waste can be used economically to raise steam or to produce hot water for use in the plant. The utilization of the excess energy contained in certain industrial wastes, e.g., waste solvents, also can be used effectively in fluidized-bed systems to provide the energy required for the disposal of a second industrial waste whose heating value is not sufficient to maintain an autogenous combustion operation.

The examples of commercial applications of fluidized-bed combustion of industrial wastes cited in this chapter all involve the utilization of energy for the disposal of the waste itself and/or another waste. The one example in which a waste heat boiler is used to reclaim heat for steam generation demonstrates the energy recovery feature.

IV. APPLICATION OF FLUIDIZED-BED TECHNOLOGY

A. Waste Liquor from Neutral Sulfite Semichemical Pulping Operations

The use of fluidized-bed combustion for the treatment of sodium-based neutral sulfite semichemical (NSSC) pulping waste liquors is commercially practiced. In NSSC mills, a pulping liquor containing sodium sulfide (NA_2SO_3) and sodium carbonate (Na_2CO_3) is used as the pulping medium. During pulping, lignin and other organic matter are extracted from the wood and the spent pulping liquor contains sodium–sulfur compounds of varying degrees of oxidation in association with the organic matter extracted from the wood. For the most part, these compounds are in solution in the waste liquor, although there may be a small quantity of suspended solids, such as cellulose fibers. The total solids in the spent pulping liquor have a gross heating value of about 5500 BTU/lb (dry), and contain about 30% carbon and about 3% hydrogen.

The spent liquor from the pulp washing operation contains about 10% total solids. To support autogenous combustion of the liquor in a fluidized-bed furnace, these waste liquors must be concentrated to about 35% total solids. Conventional incinerators burning pulp mill waste liquors require liquor concentrates of 55 to 70% solids. Concentration is normally done in three-body, triple-effect evaporators having a steam economy of about 2.5. The concentrated liquor is pumped to a storage facility before introduction into the fluidized-bed system.

The temperature of the fluidized-bed furnace is maintained at about 1325°F. The concentrated liquor is introduced as a liquid dispersion into the freeboard area of the furnace. This method of feeding has several advantages over introduction of the liquid feed below the level of the fluidized bed. As the liquid feed contacts the heated exhaust gases rising from the fluidized bed, part of the water content of the feed is evaporated and the remainder of the feed drops into the bed where the organic portion is oxidized to carbon dioxide and water vapor and the residual inorganic chemicals are oxidized to their stable oxidation state, i.e., sodium sulfate and sodium carbonate. Oxidation of the organic matter provides sufficient thermal energy to maintain the reaction temperature constant and at the desired value. The production of inorganic chemicals, sodium sulfate and sodium carbonate, is accomplished. They deposit in the bed and represent material recovery possibilities of the fluidized-bed combustion. They agglomerate and are discharged as nearly perfect spheres within the size range of 10 to 65 mesh as shown in Fig. 4.

The exhaust gases, containing some entrained particulate material, are passed through a cyclone separator where the particulate material is separated from the exhaust gases and returned through a conveying mechanism directly to the fluidized bed. The exhaust gases are then treated in a wet scrubber circuit to remove any entrained material, and to recover some of the sensible heat contained in the exhaust gases. The sensible heat can then be used in the plant. Either water or weak waste pulping liquor can be used as the scrubbing medium.

Perhaps one of the most important characteristics of fluidized-bed incineration systems is the control of the composition of the exhaust gases.

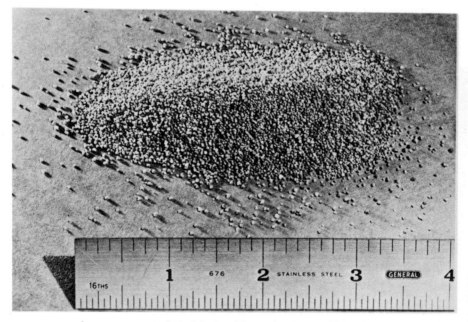

Fig. 4. Typical Na_2SO_4–Na_2CO_3 product from NSSC unit.

TABLE I

Typical Analysis of Exhaust Gases from Fluidized-Bed Treatment of NSSC Pulping Waste[a]

Component	Volume %	Component	Volume %
CO_2	12.3	Methyl mercaptan	Nil
O_2	6.7	Disulfide as C_3H_8SH	Nil
N_2	79.2	Carbonyl sulfide	Nil
H_2	0.5	H_2S	Nil
CH_4	0.3	SO_2	Nil
C_2H_6	0.03		
C_2H_4	0.02		
C_2H_2	Nil		
CH_3OH	Nil		

[a] By mass spectrograph.

In the commercial units used by the pulp and paper industry for combustion of waste pulping liquors, the exhaust gas units are monitored for the presence of offensive gaseous compounds which could cause air pollution problems—the presence of gaseous sulfur compounds would be objectionable from an air pollution standpoint. A typical analysis of the exhaust gases from one of these units is shown in Table I. As can be seen, the effluent gases do not contain any sulfur compounds that can contribute to air polution.

B. Pulping Operations

Following the installation and successful operation of the first commercial fluidized-bed unit for processing NSSC pulping waste effluent, other possible uses for this technique in the pulp and paper industry were considered. The relative simplicity of the recovery and regeneration of cooking liquor from magnesia-base pulping effluent made this system an attractive candidate for research and development efforts. Basically, the magnesia-base recovery system involves the combustion of the organic components of the waste liquor under conditions causing evolution of the sulfur as sulfur dioxide and the production of the magnesium as highly reactive magnesium oxide. This latter product is readily available for reaction with sulfur dioxide to regenerate pulping liquor, by their recombination, according to the reaction

$$MgO + H_2O + 2SO_2 \rightarrow Mg(HSO_3)_2$$

Preliminary experimentation on the adaptation of the Copeland Process to the treatment of magnesia-base pulping waste effluents was conducted at

Battelle's Columbus Laboratories early in 1964 under the sponsorship of the Container Corporation of America. The results of the preliminary study indicated that the organic content of the concentrated magnesia-base waste liquor could be burned in a fluidized-bed system and that a product having a magnesia availability in excess of 95% could be prepared. The characteristics of this waste are shown in Table II.

Experimental data which were generated during operation of the pilot-scale fluidized-bed unit indicated that 98% of the magnesia fed into the unit was collected either in the bed or as a finely divided cyclone product. Over 96% of the magnesia was determined by a standard analytical technique to be available for reaction with sulfur dioxide. These results predicated good magnesia recovery in a commercial system utilizing this technique for treating magnesia-base waste liquor. The experimental data also showed that over 95% of the sulfur was eliminated as sulfur dioxide during the fluidized-bed incineration of the red liquor.

The following conclusions were made regarding the fluidized-bed incineration of magnesia-base pulping waste effluent:

(1) The fluidized-bed system can be operated successfully at temperatures ranging from about 1500 to 1750°F, while burning the organic content of the waste liquor. (From the standpoint of heat recovery, the higher end of the temperature range is more desirable.)

(2) Autogenous operation can be attained in the pilot-scale unit with a feed liquid solids content of 45 to 50%. (Autogeneous operation in larger installations with more dilute solutions should be readily attainable on the basis of past experience with other waste liquors.)

TABLE II

Characteristics of Concentrated Magnesia-Base Waste Liquor

Specific gravity	1.281
Solids, oven dry (%)[a]	50
Chemical analysis of liquor (gm/liter)	
MgO	49.9
Total S	54.2
Chemical analysis of oven dry solids	
Carbon (%)	40
Hydrogen (%)	4.9
Heating value (BTU/lb)[b]	7147

[a] Dried at 250°F.
[b] By bomb calorimetry.

(3) The fluidized-bed unit can be operated successfully to produce a finely divided magnesia product in which over 90% of the MgO is available for reaction with an aqueous solution of sulfur dioxide.

(4) Magnesia recoveries in excess of 95% of that present in the waste liquor should be attainable in a commercial unit—based on the recovery of 98% of that fed into the pilot-scale unit in the various product streams.

(5) Recovery of 90 to 95% of the sulfur dioxide from gas streams containing about 1% of sulfur dioxide can readily be accomplished to produce an acceptable cooking liquor.

On the basis of both the pilot-scale data from the study of the incineration of magnesia-base liquor and the operational data from existing NSSC installations, an integrated commercial-scale fluidized-bed system was designed and constructed at the Wausau Paper Company mill in Brokaw, Wisconsin, in the late 1960s. A sketch of this system is shown in Fig. 5. This unit has operated successfully since that time.

The fluidized-bed combustion system designed and constructed for the treatment of magnesia-base waste pulping liquor differs from that used for treating NSSC waste in that it includes a cooking liquor regeneration system and a waste heat boiler for generating steam. A waste heat boiler is used primarily because the bed can be operated at higher temperatures ($\sim 1700°F$) than the bed in a unit in which NSSC waste liquor is being burned. These additions ensure the most effective conservation of both chemical components and the fuel value of the waste liquor.

Energy recovery in the pulp and paper industry is widely accepted and growing. In the late 1960s, the use of residues supplied 37% of that industry's fuel requirement. Today it supplies 45% of the energy needs of papermakers. Fluidized bed combustion offers that industry a most useful technique for waste-to-energy conversion. Fluidized beds are useful in other industries also.

C. Petroleum Refining Wastes

Fluidized-bed combustion has also been applied to the disposal of oily sludges and spent caustic produced in petroleum refining operations. A commercial-scale demonstration unit for this purpose which was installed at the American Oil Company plant in Mandan, North Dakota went on-stream in June 1969. The feed to this incinerator consists of oily sludges containing up to 50% water and 5% solids, typically API separator sludge; tank bottoms of various types from oily sediments to waxy cakes; stable emulsions from processes and from slop recovery operations; and spent caustic solutions, typically 2 to 8% caustic, rich in mercaptans and sulfides.

206

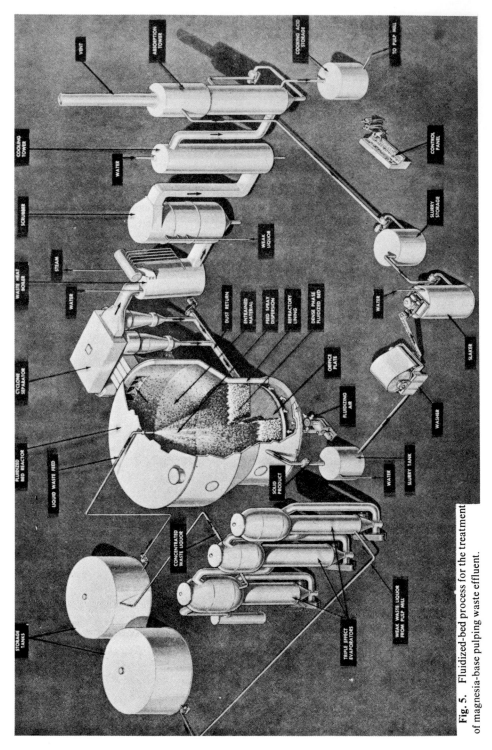

VENT

ABSORPTION TOWER

COOLING ACID STORAGE

TO PULP MILL

COOLING TOWER

WATER

SCRUBBER

WASTE HEAT BOILER

STEAM

WATER

WEAK LIQUOR

CONTROL PANEL

SLURRY STORAGE

DUST RETURN

ENTRAINED MATERIAL

FEED SPRAY DISPERSION

REFRACTORY LINING

DENSE PHASE FLUIDIZED BED

CYCLONE SEPARATOR

ORIFICE PLATE

WATER

SLAKER

FLUIDIZED BED REACTOR

LIQUID WASTE FEED

FLUIDIZING AIR

WASHER

SOLID PRODUCT

WATER

SLURRY TANK

STORAGE TANKS

CONCENTRATED WASTE LIQUOR

TRIPLE EFFECT EVAPORATORS

WEAK WASTE LIQUOR FROM PULP MILL

Fig. 5. Fluidized-bed process for the treatment of magnesia-base pulping waste effluent.

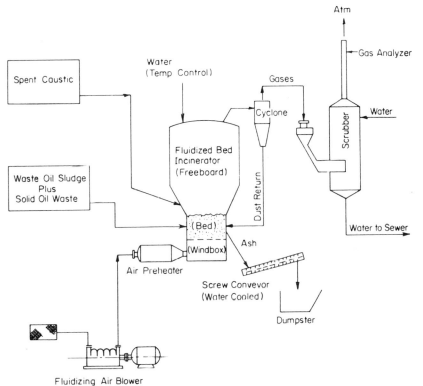

Fig. 6. Typical fluidized-bed refinery waste incineration unit.

A sketch of the Mandan incineration system is shown in Fig. 6. This unit was designed to process about three barrels of oily sludge and two barrels of spent caustic per hour.

Although the refinery wastes contain sulfur in several forms, the exhaust gases from the Mandan incinerator contain less than 0.8 ppm of sulfur dioxide. This again illustrates the capability of fluidized-bed systems to minimize pollution of the air with sulfur dioxide by capturing this component in the bed provided sufficient alkaline materials are present.

D. Petrochemical Wastes

Over the years several fluidized-bed combustion systems have been designed and constructed for the disposal of sludge from sewage plants. Somewhat similar sludges are produced in the biological treatment of various organic wastes and the feasibility of using a fluidized-bed system for

TABLE III

Simulated Petrochemical Sludge

Material	Percent
Volatile matter	12.80
Calcium carbonate	5.59
Sodium chloride	2.53
Calcium chloride	0.15
Calcium sulfate (anhydrous)	0.03
Water plus inert materials, etc.	79.90
Total	100.00

the disposal of such sludges was explored in the late 1960s. Following a pilot-scale study in which simulated sludge was used as a feed material, a commercial-scale unit was designed and constructed for burning sludges produced during the biological treatment of wastes from a petrochemical complex. The composition of the simulated sludge is shown in Table III.

Among the complicating factors involved in the combustion of this and many other organic wastes is the presence of several inorganic salts which form ternary and quaternary mixtures. Melting points encountered in such systems may be significantly lower than those of the individual pure salts and the residual material often fuses at temperatures below that at which complete combustion of the organic constituents occurs. In a fluidized-bed system such fusion leads to defluidization of the bed and makes the system inoperable. Although ternary and quaternary systems of calcium and sodium salts were present in the simulated sludge used in the pilot-scale experimentation, they did not cause any significant problem at the temperatures required for complete combustion of the organic materials, i.e., 1300 and 1500°F.

The commercial-scale system designed for the disposal of petrochemical sludges involved the use of a thickener and centrifuges to dewater the excess sludge to about 12 to 18% of solids. The dewatered sludge is introduced into the freeboard of the fluidized-bed combustion unit and is burned in the bed at a rate of about 6 tons/day. Because of the relatively low solids content of the sludge, supplemental fuel is required.

The residual solids are collected in the bed under conditions which favor their agglomeration. The nominal superficial gas velocity for this unit is about 3 ft/sec.

Among the factors involved in the selection of a fluidized-bed combustion system for this particular application was the ability to attain practically complete destruction of the organic components of the sludge without the creation of odor problems associated with many of the components which may be present. Other factors involved in the selection include the ease of operability of the system, low capital and maintenance costs, and the dependability of the system.

V. CONCLUSION

A few illustrations of the successful commercial-scale application of fluidized-bed technology to the conservation and utilization of energy from industrial wastes have been discussed in this chapter. Additional potential uses of this technique lie in the combustion of waterborne, semisolid, and solid wastes from such industries as plastics, paint, and rubber. Preliminary mini-pilot-plant experimentation has already been conducted with some of these wastes and their combustion in a fluidized-bed system appears to be both technically and economically feasible.* Other wastes which are or may be amenable to such treatment include those from the pharmaceutical industry and waste wood and bark from the paper and lumber industries.

Among the conclusions which can be drawn from both commercial and experimental practice is that the heating value of selected industrial wastes can be utilized most effectively when the energy is used for the direct treatment of the waste itself, i.e., the evaporating of water contained in the waste material. A second effective use of such energy is the treatment of a second waste stream which requires the evaporation of water.

It is concluded from this experience that fluidized-bed combustion is an effective technique for utilizing energy contained in many dilute industrial wastes. In some cases where the properties of the residual inorganics will permit operation at reasonably high temperatures, additional energy can be recovered via waste heat boilers. This technique has also proven to be extremely effective for the disposal of wastes under conditions which eliminate the production of various odors and other air pollutants through the close control of temperature which it affords. The continued development of additional uses for fluidized combustion for the utilization of energy contained in industrial carbonaceous wastes seems assured.

* "Fluidized-Bed Incineration of Selected Carbonaceous Industrial Wastes," U.S. Environmental Protection Agency Grant No. 12120 FYF, March 1972.

Chapter XIII

NONTECHNICAL ISSUES IN THE PRODUCTION OF FUELS FROM WASTE

David A. Tillman
MATERIALS ASSOCIATES, INC.
WASHINGTON, D.C.

I. INTRODUCTION

The production and utilization of fuels from municipal, industrial, agricultural, and silvicultural wastes depend not only on technical design but also on the solution to pressing nontechnical problems. These nontechnical issues include gathering waste in an economically viable manner, marketing the fuel produced, and the vexing complications of financing the plant. These problems cut across all forms of waste.

Other nontechnical issues have been raised including tax treatment of residual fuels versus virgin fuels (i.e., coal or petroleum), possible freight rate discrimination in the transportation of recycled products, and similar debatable points. Although of legitimate concern, these issues are of less significance than waste accumulation, fuel marketing, and plant financing. Those fundamental three problems are basic concerns in the establishment of a waste-to-fuel installation.

II. FEEDSTOCK ACCUMULATION

Both municipal and nonmunicipal waste-to-fuel plants rely on the gathering of significant volumes of waste for processing. Industrial plants using agricultural or silvicultural waste face similar economies of scale problems. The optimum volume of waste to be received and processed daily is a function of the trade-off between plant operations and waste transportation. This principle holds regardless of the type of waste being processed.

For municipal waste disposal facilities, the minimum economic size appears to be 1000 tons/day. Beyond that point the trade-off problem arises.

Figure 1 depicts a simplified theoretical region for municipal waste generation. It has a central city of 350,000, a ring of five towns with 100,000 population each, a second ring of seven towns with 50,000 residents each, and a third tier of towns with 20,000 people in each of the 15 communities. The region generates 3 lb of waste per person per day. This simple illustration, which assumes a possible resource recovery plant in the center of the city, highlights the trade-off.

From the point of view of municipal solid waste availability, the central city does not generate enough solid waste to support a recycling system or a major landfill. By including the five towns immediately surrounding the central city, however, some 1275 tons of municipal waste will be available for processesing on a daily basis. Thus the central city, and the immediately surrounding communities, can produce an economically feasible project. If, however, the seven second-tier communities are included in the region, waste generation increases by another 525 tons/day; the system can be designed to handle 1800 tons/day of municipal waste. The 15 communities in the third tier add only 450 tons of waste per day, driving production up to 2250 tons/day. From the point of view of waste generation, then, optimal incremental additions of waste appear to come from the central city and the first two tiers of surrounding communities.

From the point of view of collection transportation, the central city obviously enjoys an extreme advantage. If the daily tonnage generated by each community is divided by the distance to the recycling center, then the following ratios appear: central city, 140 tons/mile; first tier communities, 12 tons/mile; second tier communities, 3.33 tons/mile; third tier commu-

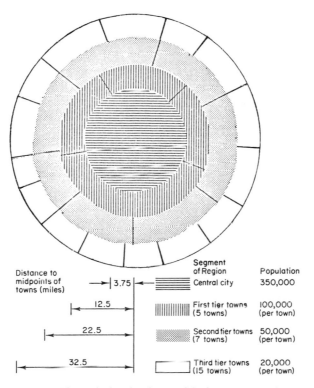

Fig. 1.   Theoretical region for municipal waste generation.

nities, 0.93 tons/mile. Table I presents the economic trade-off factors. Clearly, the transportation cost to the recycling facility remains acceptable in the central city and the first tier communities. The second tier communities have a marginally acceptable transportation cost; the third tier communities remain outside the effective transportation radius.

The problems for wood waste and animal waste are little different. If one company generates a sufficient volume to justify energy recovery facilities, then no problem is encountered. That situation is analogous to a central city generating all the waste feedstock for a MSW processing plant. If a single wood processing firm or cattle feedlot generates less than enough waste, problems multiply rapidly. The absence of wood waste utilization plants (e.g. particle board plants) in northern New England stems from the fact that all the sawmills and planing mills in New Hampshire and Vermont, together, generate only about 1500 tons/day of usable residue [1]. In the animal waste area, large feedlots have been the prime areas of consideration to date. They generate the volumes of feedstock required for efficient operation. The graph shown in Fig. 2, developed by Mitre Corporation for the state

TABLE I

The Impact of Adding Towns to a Regional Solid Waste Management System Based on the Theoretical Region Presented

Regional size	Total population	Added population	Total tons of waste generated daily in region covered	Additional tons MSW generated daily by additional towns	Average tons hauled per mile (to processing plant)	Average tons hauled per mile (to processing plant) for added MSW
Central city	350,000	n/a[a]	525	n/a	140	n/a
Central city plus first tier towns	850,000	500,000	1275	750	20.4	12
Central city and first tier plus second tier towns	1,200,000	350,000	1800	525	8.3	3.33
Central city and first two tiers of towns plus third tier towns[b]	1,500,000	300,000	2250	450	3.2	0.93

[a] n/a means not applicable.
[b] By the time the third tier of towns is added, the increased costs of transporting the waste to the processing plant may more than offset the economies of scale generated by the additional tonnages of waste available for processing.

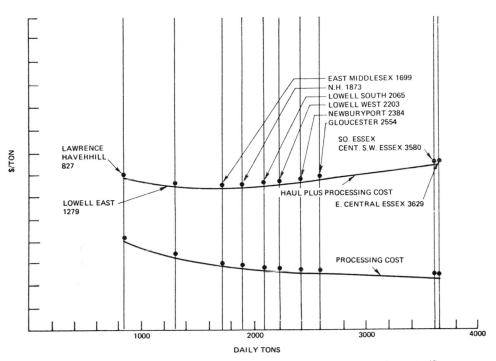

Fig. 2. Economics of scale in gas pyrolysis (MITRE preliminary estimates). (Source: Mitre Corporation.)

of Massachusetts, summarizes the problem, showing decreasing plant costs being overcome, eventually, by increased waste gathering expenses. This calculation includes the use of transfer stations to increase the effective collection radius.

III. PRODUCT MARKETING

Three issues appear in product marketing: (1) the nature of the fuel and its potential synergy with currently employed commercial fuels, (2) the structural relationship between the fuel producer and fuel user, and (3) the physical proximity of the fuel producer to the fuel user. The first question provides the interrelationship between technical and nontechnical issues. The second and third questions affect plant ownership and financing.

A. The Nature of the Fuel Produced

Fuels produced from waste must be compatible with existing infrastructural arrangements. Since waste-based fuels are, at best, a supplementary source of energy for society, they cannot command their own combustion systems on a widespread basis. The currently available and contemplated fuels resulting from the conversion of waste include shredded solids, heavy liquids, light liquids (methanol), low-BTU gas (100–200 BTU/ft³), medium-BTU gas (300–400 BTU/ft³), methane, and steam. Each fuel has been explored, technically, in one or more of the preceding chapters.

Significantly, only steam and methane are virtually interchangeable with existing fuels. Thus incineration of municipal waste, accompanied by steam recovery, has become accepted in a variety of cities as was shown in Chapter II. Similarly, fluidized-bed combustion of papermaking wastes—pulping liquors—has been increasingly useful in that industry. Methane production is less well advanced, but the interchangeability of this product from municipal waste and feedlot waste with natural gas provides a significant economic incentive for the Syngas process and the anaerobic digestion of feedlot waste. Methanol, a well-understood fuel, comes close to steam and methane in institutional acceptance.

Other fuels—shredded solids, heavy liquids, and low- and medium-BTU gases—have varying degrees of compatibility with existing commercial fuels. It is significant to observe that, whereas steam and methane can be used ubiquitously, the other fuels must find their way into utility and industrial markets. Further, the decreased BTU content per unit of fuel has a significant impact upon the price which these energy forms can command. Thus, these fuels are basically confined either to electric power generation (i.e., St. Louis, Missouri) or manufacturing operations.

B. Fuel Producer and User Structural Relationships

Fuels may be produced and used by the same economic entity. They may be produced by one organization and sold to another economic organization. Those two options exist. The difference is significant in terms of the project acceptance. The difference becomes even more significant when one contemplates the growth of this industry.

If the fuel is produced and used by the same economic entity, projects may move faster than if fuel product sales are involved. Accommodation can be made for the lower economic value of the fuel—and appropriate credits and debits can be entered. Thus this approach has become the most widely utilized approach to date.

When Ames, Iowa decided to enter the fuel-from-waste business, the municipality built the resource recovery plant and "sold" the fuel product to a municipal power plant. The same economic entity which produces the waste-based fuel also uses that fuel. The resource recovery plant prices its fuel, to the utility, on a BTU basis. It charges as much for refuse-derived solid fuel as the utility pays for coal. The economic advantage, which ranges from 10 to 20% of the price of the fuel sold, is obvious.

In the use of fuels from agricultural and silvicultural wastes, this has been the overwhelming approach. The 1.0 quadrillion BTU of energy generated and used by the pulp and paper industry—turning waste into fuel—is built on this approach. Black liquor, bark, and sawdust are burned to produce steam. The resulting steam is then used in the papermaking complex. One significant exception is in Eugene, Oregon, where hogged fuel is purchased and then burned to produce steam and electricity for a municipally operated utility. There the wood waste is sold to the energy customer.

The option used in Eugene, Oregon is to produce a fuel and sell it on the energy market. This option has been used in the municipal waste treatment industry. It will become essential in the production of methane from feedlot waste or methanol from wood and industrial waste.

In the produce-and-sell mode of operation, the producer must provide some incentive to the customer for the purchase of the fuel. The methane producer can offer an obvious advantage—the filling of pipelines. Domestic production of natural gas has peaked and is declining at a constant and apparently irreversible rate. For the other fuel producers, however, the incentives must be more direct.

The practice of producing fuel and selling that fuel to another user is becoming more and more prevalent as municipal systems come on-line. In these systems, the fuel price per BTU is lower than the fuel price of the natural fuel counterpart. This has been the consistent trend. These are "off spec" fuels. They do not offer the same quality as naturally occurring fuels such as coal or petroleum. Thus, the fuel producer provides the user with an economic incentive to make a shift. Typically that incentive is in the 15 to 25% range, when the recycled fuel is compared to natural fuels.

C. Physical Relationships between Producers and Users

Proximity to fuel markets is the final marketing criterion. Optimally, the plant should be located directly adjacent to the fuel market, since fuel products constitute 80% of the output of a resource recovery facility. Three examples of this optimal location can be cited.

In St. Louis, the two facilities to be owned and operated by Union Electric will be sited directly adjacent to the Meramec and Labadie power plants. No transportation cost will be incurred in moving the dry shredded fuel from the resource recovery plant to the fuel user or power plant.

In Saugus, Massachusetts, the RESCO incinerator is located directly across the river from the General Electric Lynn Works. A pipeline, shown

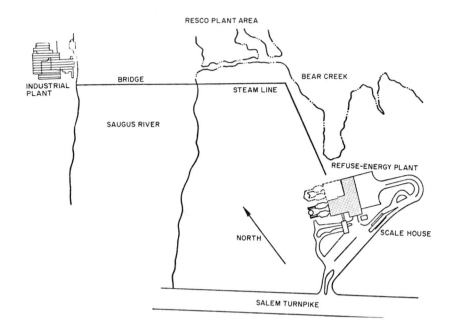

Fig. 3. Minimizing the distance between the resource recovery plant and the market for the energy product is one of the most important aspects of marketing. The short distance between the RESCO incinerator in Saugus, Massachusetts and the General Electric manufacturing plant in Lynn, Massachusetts demonstrates one optimal situation. (Source: Wheelabrator–Frye Company.)

in Fig. 3, moves the steam from the Saugus resource recovery facility to the fuel user, the jet engine manufacturing facility of General Electric. Again, no appreciable operating costs are incurred in moving the energy product from the producer to the user.

In Ames, Iowa, the resource recovery facility is located three blocks away from the central business district, and 800 ft away from the municipally owned power plant. Thus, it can deliver the fuel fraction by pneumatic piping from the resource recovery facility to the fuel product consumer.

For the directly substitutable fuels, proximity must be to a distribution system (i.e., a local steam or natural gas pipeline), but none of the off-spec fuels can be moved significant distances. For the lower grade fuel, proximity must be to the end user itself.

IV. FINANCING FUELS FROM WASTE

Financing is the final consideration of a nontechnical nature. For municipal systems this consideration carries the question of public or private ownership of the system. Figure 4 is a relevance tree spelling out the ownership–financing mechanism options. Thus, the left-hand side applies only to municipal solid waste systems whereas the right-hand side relates to all types of waste-to-fuel projects.

A. Public Ownership Financing

Certain general advantages apply to public ownership options. Communities maintain their traditional responsibility for waste disposal; further, they maintain direct quality control over that operation. A second advantage may accrue to the community through lower cost capital from the tax-free bond market.

Disadvantages also exist for public ownership. Public ownership impacts upon the financial strength and credit rating of a community. Public ownership puts the community into the business arena as a seller of manufactured products. As a rule, communities lack this expertise. Marketing, the critical function of private enterprise, is not a traditional activity of communities. The third disadvantage of public ownership is this: The facility will bring no tax revenues to the community. Thus, the facility will not be a contributor to the overall economic base of the community. Such implications must be seriously considered before a community moves to secure financing of a general obligation bond or revenue bond nature. Both types of bonds merit review.

1. General Obligation Bonds   General obligation bonds may be used to finance resource recovery projects. Ames, Iowa employed such bonds to finance its 200-ton/day dry mechanical system which produces fuel and metals. This mechanism remains the traditional approach for municipal financing. When communities use this mechanism, they develop the project and its costs. These capital requirements are lumped with other current city needs. Then the community holds an election, or referendum, to determine

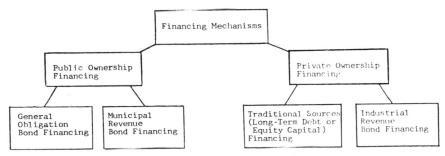

Fig. 4.   Relevance tree of financing.

whether or not the taxpayers support the issue. Finally, the community brings that bond issue to the tax-free municipal bond market.

When the bonds reach the marketplace, they are sold on the basis of the community's credit rating, rather than on the merits or demerits of the specific project. Thus, the community does not have to prove the wisdom or feasibility of the specific recycling project to the prospective purchasers. Rather, the community develops a prospectus which discusses the project, demonstrates the legality of the issue through bond counsel opinion, and demonstrates the financial responsibility of the city or region (county) to pay off the obligations.

The advantages which accrue to users of this approach include not only the general advantages of municipal ownership but also the possibility of low interest rates. On its issue of $5.6 million, Ames, Iowa gained an interest rate of 5.3% for 20 years. Such interest rates in a tight capital market appear extremely attractive. Because Ames, Iowa owned and operated its own electrical generating station, the boiler modifications of $170,000 were financed in this bond issue.

Disadvantages also characterize this approach. The use of government general obligation bonds impedes the ability of a community to build additional facilities, such as schools and firehouses. Thus, before a community chooses the G.O. bond route, it must evaluate the capital needs of the community.

Further, the G.O. bond route may not provide capital. When a community goes to the taxpayers with the referendum, the residents of the community or region may be voting upon their overall impression of the local administration as well as the specific project involved. In 1975, the city of New Orleans conducted a referendum for general obligation bonds to pay for a new sanitary landfill facility. In the wake of general criticisms, this $40 million bond offering was voted down. This is not an isolated example.

Additionally, communities with marginal credit ratings may not get money at low interest through this mechanism. The agonies of New York City during the summer and fall of 1975 blew serious winds of caution into the municipal bond market. On July 30, Municipal Assistance Corporation bonds offered by the State of New York to assist New York City carried an interest rate of 9½% [2].

Cities generating fuels from waste may have further problems employing municipal bonds of a general obligation type due to their general inapplicability to regional plants. In a regional situation, unless special circumstances exist, one community must accept the entire responsibility for capital formation. In such regional cases, the most probable situation would be that the central city accepts the capital raising responsibility. Normally, however, the central city faces the largest demands for capital from competing programs. In most regional situations, unless special legislation provides for tax base sharing, general obligation bonds create an untenable situation, provoking serious strains between the central city and the satellite suburbs.

2. Municipal Revenue Bonds Municipal revenue bonds emerge as a variation on the public financing scene. Like the general obligation bonds, revenue bonds are tax exempt. Thus, they are highly attractive to investors seeking tax shelters. If the community seeks this alternative, it must do so on a project-by-project basis.

Communities and regions evaluating the municipal revenue bond alternative begin with the technical selection process and an analysis of the technical feasibility of the project. At such time, communities must either obtain contracts for the products or commit other sources of revenue to the project. If they are to be applied to the municipal revenue bond repayment, contracts for the sale of energy must be for the life of the bonds. Additional sources of revenue can include a tax on refuse produced.

The structure of municipal revenue bonds varied significantly from the structure of general obligation bonds. Revenue bonds must maintain a debt service reserve; however, they do not require an election or referendum. When the prospectus reaches the bond market, it is judged on the merits of the project itself; it is not judged on the merits of the city's general credit rating.

Advantages accrue to communities opting for public ownership with financing by municipal revenue bonds. Revenue bonds do not tie up the bonding capacity of the city. They do not impede the ability of a community to finance other capital projects.

Disadvantages also accrue to communities using this system. Not many cities have uncommitted revenues; thus, some must increase taxes in the form of a refuse tax. Further, if the project does not have a good financial rating, it may not be funded by the bond market. Thus, if the project achieves an "A" or better rating, it may develop reasonably low interest rates. If it achieves a "Baa" rating or lower, however, it may not get monies. The July 30, 1975, $3.5 million offering by Buffalo, New York, for sewage treatment facilities, rated Baa, had no takers [2].

B. Private Ownership and Financing

Waste-to-fuel projects of a nonmunicipal nature come from the private sector. These facilities, plus several municipal waste-to-fuel projects, have been designed to be owned and operated by private industry. Corporations gain advantages from private ownership and operation of these resource recovery facilities, in addition to an assured source of fuel. Such advantages include investment tax credits, accelerated depreciation, and other financial incentives. They provide a return on the dollar invested to those corporations with vision, and with solutions to their own or municipal waste disposal problems.

The principal disadvantage associated with corporate ownership and corporate financing is the fact that corporate monies are more expensive than municipal monies. With the exception of industrial revenue bonds, corporate financing is not tax exempt. An additional disadvantage in employing corporations comes through the issue of risk taking. The corporation, which must either invest its equity or gain financing from high cost corporate bonds, must also take the entire financial and technological risk associated with the project if it owns and operates the solid waste disposal facility. Because the corporation assumes the risks, it must expect a reward. That reward comes from profits which are at levels above the return on investment for secured investments.

Corporations have two basic sources of funds: traditional and industrial revenue bonds. Both merit cursory examination.

1.   Traditional Corporate Financing   The waste-to-energy facility which is privately owned and operated may be financed through the traditional sources of capital. The corporation has at its disposal internal financing mechanisms which consist of reinvesting retained earnings. In addition, a corporation may go to the stock market or the corporate bond market.

If the system handles municipal waste, the private corporation must obtain contracts for the disposal of municipal waste. These contracts are the "put or pay" contracts. In such a contract, communities guarantee to supply the facility with a guaranteed minimum tonnage of solid waste. The communities also guarantee to pay a fixed cost per ton for that delivered (or undelivered) municipal waste. In Saugus, Massachusetts, the nine initial communities together guaranteed RESCO some 620 tons of municipal waste per day, at a price of $13 per ton [3]. In a nonmunicipal waste plant, where the plant handles the industry's own waste, this contracting is avoided.

In addition to the income derived from waste disposal, municipal energy recover plants must have "take or pay" contracts for the sale of fuels produced. Such contracts read that if the facility is built, the customer will take the fuel produced and will pay an established cost for it. If an industry uses the fuel itself, particularly if its own waste is being converted into fuel, this is again avoided.

After these contracts have been negotiated, the corporation approaches traditional sources of financing. The corporate board of directors must decide if a corporation is to commit between $10 and $100 million of its own capital to the project. The corporate board cannot favor such a project unless its return on the investment is significant.

When the corporation faces the possibility of technical risks, it seeks a higher return on its investment. Risk taking is a cost item in the total project planning. If the board of directors of any corporation is assured that it can gain an acceptable return on its investment, then it will accept such a risk. It will finance the project out of current revenues or other traditional corporate sources of income.

Corporations gain tax advantages from ownership. Two tax advantages— the investment tax credit and accelerated depreciation—can be used either as a method for reducing the effective equity in the plant or increasing the return on investment during the first year of the project's operation.

Consider the theoretical example, shown in Table II, of a private corporation investing $35 million in a municipal waste-to-fuel facility. Of the $35 million, this corporation borrowed $25 million and invested $10 million of its own internally generated cash. In the first year, the corporation gained a 10% investment tax credit of $3.5 million. By employing accelerated depreciation over a 15-year period, the corporation also gained $4.67 million of depreciation money. If this corporation borrowed money at 10%, an additional $2.5 million deduction could be achieved. From the

TABLE II

First Year Tax Advantages of Private Ownership and Financing [a]

Conditions	
Total investment	35,000,000
Equity investment	10,000,000
Long-term debt (10% money)	25,000,000
Tax advantages	
Investment tax credit (10%)	3,500,000
Accelerated depreciation (15-year term)	4,670,000
Interest deduction	2,500,000
Cash returned to owner	
Investment tax credit	3,500,000
50% of tax deductions	3,600,000
Accelerated depreciation	
Interest deduction	
Total	7,100,000
Net equity in resource recovery plant after first year	2,900,000

[a] In dollars.

depreciation and the deduction of interest for taxes, a total of $7.17 million in tax deductions is realized. These translate, normally, into $3.6 million of cash. Additionally, the investment tax credit provides $3.5 million in cash. Thus, in the first year, this corporation can reduce its cash investment from $10 million to $2.9 million. The advantage of ownership does provide financial incentives which merit consideration.

Certain disadvantages characterize these projects. These disadvantages include the high cost of corporate monies affecting the front-end costs and the processing cost per ton. If municipal bonds can be sold at an interest rate of 6%, and if corporate bonds are sold at an interest rate of 9%, a significant financial disincentive occurs. Despite the fact that the corporation can deduct interest payments from its taxes, the high cost of money and its impact on the corporation's ability to borrow for other projects transfer into an increased cost of waste disposal.

2. Industrial Revenue Bonds A compromise exists between strictly municipal financing and strictly private financing. This compromise comes in the form of industrial revenue bonds. Some 36 states offer tax-free financing of industrial revenue bonds and pollution control bonds to industries.

Section 103 of the Internal Revenue Code establishes the rules for industrial revenue bonds. Up to $1 million worth of industrial revenue bonds may be raised for any purpose. Further, $5 million may be raised if the corporation using the bond proceeds does not spend that amount in the three years preceding the issue or the three years succeeding the issue of the bonds. Finally, exemptions exist which eliminate any dollar limit on industrial revenue bond issues. The most well-known exemption is the Pollution Control Bond exemption, which can be used to finance air and water pollution abatement equipment, including industrial waste-to-fuel projects. Succeeding the establishment of that exemption is Section 103(c)(4)(e) of the Internal Revenue Code, which established a similar exemption for municipal solid waste facilities.

In the *Federal Register* (**40**, No. 120, June 30, 1975), this exemption was detailed. Under Definitions, these rules state: "The term 'solid waste disposal facilities' means any property or portion thereof used for the collection, storage, treatment, utilization, processing, or final disposal of solid waste."

The text proceeds to amplify what is covered for the resource recovery facility. In general it states: "Where materials or heat are recovered, the waste disposal function includes the processing of such materials or heat which occurs in order to put them into the form in which the materials or heat are in fact sold or used, but does not include further processing which converts the materials or heat into other products." That IRS ruling states that the resource recovery facility itself is covered by the bonds, but the transportation equipment to convey the fuel product—dry fuel, steam, pyrolytic gas, etc.—and all equipment after the transportation equipment cannot be covered by such bonds. In Saugus, RESCO employed this approach. Because the financial application was to the bond market, contracts were developed in detail. Contracts for put-or-pay depositing of municipal waste and contracts for the sale of products from the project were necessarily effective for the life of the bonds.

Although the community is the nominal owner of the facility and leases it to the company for bond payments, the company effectively owns the resource recovery plant and guarantees the bonds. Thus, it gains the tax advantages of accelerated depreciation and investment tax credit. Because the municipality is the nominal owner and the bonds are tax exempt, the interest rates are from 1 to 3% lower than the interest rates from traditional corporate financing.

Corporations, then, can use this approach as an alternative to traditional financial sources. The money costs less. The other requirements remain virtually unchanged. Such corporations can succeed, financially, on a slightly lower charge for waste disposal.

V. CONCLUSION

Among the many nontechnical issues, the problems of waste collection, fuel marketing, and ownership–financing remain most critical. Among these, the primary problem is financing. Fuels from waste plants, using agricultural, silvicultural, or municipal wastes, must obtain enough material to operate economically. This material must come in a relatively even flow. Second, these plants must produce a fuel which industry, if not the body politic, can use. Finally, they must produce a fuel in a facility which can be financed and operated in a manner competitive with other pollution control and/or energy production systems. While these facilities combine environmental protection and energy production, the combined economics must still be favorable.

For the municipal plants, these problems are more thorny than for industrial wastes. Industries entering the fuels-from-waste business normally generate enough feedstock of their own to justify a plant. Further, most such industries serve as their own market. In the case of methane from feedlot wastes, the natural gas pipelines can serve as a viable market. Finally, such industries normally have access to financing.

REFERENCES

1. D. A. Tillman, Particleboard raw material availability report. New Hampshire Office of Industrial Development, Department of Resources and Economic Development, 1970.
2. Phil Hawkins, "New York City problems are cited as Buffalo agency fails to sell securities," *Wall Street Journal*, p. 29, July 30, 1975.
3. $30,000,000. Town of Saugus, Massachusetts: solid waste disposal revenue bonds. Prospectus for bond sale published by White, Weld & Co., New York, New York, p. A-17, August, 1975.

INDEX

227